Mario Carpo
Beyond Digital

THE MIT PRESS   Cambridge, Massachusetts   London, England

Mario Carpo
Beyond Digital

Design and Automation
at the End of Modernity

1   Ways of Making

There is a story I have been telling my students for many years now, which I still find useful for introducing some core themes of my classes. More recently, when all teaching went online due to the pandemic, I had to tell this story in front of a computer screen, often prerecording it in the absence of any audience. Inevitably, under these extraordinary conditions, after a while I ended up making clips of various parts of the story and reusing them as prefabricated blocks, or "segments," in a number of recombinable, modular talks. If this may seem an abomination—and it is, both pedagogically and technically—then I guess we all do something similar in daily life; we can all think of one or two friends who always end up rehashing the same anecdotes, regardless of the story they're telling. The history of classical and modern rhetoric, in turn, offers plenty of examples of what used to be called *topoi*—originally lists of specific issues that speakers were expected to address when discussing a given subject, eventually turned into blocks of ready-made speech that the orator could cut and paste at will. In this sense, during the first months of the pandemic, video conferencing platforms like Zoom simply added more visibility to a very ancient, indeed ancestral, way of speaking.

With a significant difference though: an artisan of the spoken word, the storyteller lives in a world of embodied experience.[1] Stories that are spoken morph and change all the time. Every time a story is re-enacted—even when told by the same speaker—new details are added, and others are dropped. Teaching in a lecture theatre is not exactly, or at least not always, storytelling; but it is never far from it, either. Some teachers always tell the same story; some invent a new one every term; most have a reasonable refresh rate and bring in some new content every year—more after a sabbatical, where sabbaticals still exist. Teaching in the flesh, even when reduced to simple, monodirectional delivery (technically, broadcasting: one person speaking to many, no questions asked) always admits some degree of experiential randomness, tacit feedback (background noise, laughter) and off-the-cuff improvisation. No two classes are the same.

That was before Zoom. In the pre-digital world, communication comes first (someone speaks), notation later, or never (someone takes notes or records the voice of the speaker). In the digital world, notation comes first—as all that is digital is by definition recorded in binary form—and communication later, or never.[2] The explosion

2

of digital video as the communication technology of choice during the pandemic has converted the flow of the spoken word into snippets of exactly repeatable verbal statements—delivered once, then seen forever, and in theory everywhere. This result is not by itself unprecedented: the traditional tool to freeze the disembodied word and transmit it to generic, undifferentiated audiences across space and time used to be alphabetic writing—even more so after the invention of the printing press. By leapfrogging from orality to digiteracy, the current technologizing of the word has incidentally side-lined literacy.

There is hence something anachronistic and almost quixotic in the endeavor I am about to embark on. This introduction—the first chapter of this book—is a tale I've often told but never written, a true anecdote in the etymological sense of the term. I shall write it now for reasons that will become evident with the unfolding of this book—with apologies for the colloquial tone that will reveal the oral lineage of my story, and with apologies, more particularly, to many students around the world who will have already heard it, *viva voce* or more likely, I fear, online.

So this is what I would have told, and shown, my students in those distant times when we still had classrooms and blackboards (real ones or electronic). I would have drawn a horizontal arrow on said board and explained how that line, oriented left to right as some conventions dictate, stood for the universal chronology of humankind, from the very beginning—whenever it be—to our present. Next, I would have divided that line into three roughly equal segments: at left, the age of hand-making, which was the universal human condition before the invention of machines; in the middle, the age of mechanical machine-making, when hand tools became actual machines (I never delved into the definition of either); and at the end, the age of digital making, when machines became electronic and started to function with a new technical logic, different from and in many ways opposite to the analog logic of yesterday's mechanical or electromechanical machines. Hand-making, mechanical machine-making, digital making: all ways of making of all time and place boil down to these three technical logics or, indeed, technologies: that of the artisan, that of the factory, and that of computation. The technical logic of computation is now changing the world as deeply and irrevocably as the technical logic of mechanical machine-making

3

did at the start of the industrial revolution, but as the digital turn has only just started, we are still trying to understand what it is precisely, and how it will pan out.

These three ways of making developed sequentially over time, but the rise of each new technology never completely obliterated the preceding one—which is why we now live in a technical environment where most things are machine-made, but some are still made by hand, and a few are already made digitally (i.e., made by machines driven by computers and following their technical logic). Each way of making has its own spirit, so to speak, which partakes of the society where that mode of production took root and thrived, while leaving in turn an indelible trace on all products thus made. Look at that chair, and guess: was it made in an artisan shop? In an industrial factory? In a computational Fablab? Designers and makers may try to cheat, yet by and large one can always tell. Hence the query: what's the *sign* of craft? What's the *indexical trace* of industrial making? What does computational making *look like*? To answer these questions, we should first see what is specific to each of these three technical logics.

1.1   Hand-Making

All that is made by hand bears the trace of the free, unimpeded gesture of the hand that made it. Of course, some hand tools we wield may leave a trace too, yet regardless of the instruments involved one universal rule seems to apply to all organic movements of the human body: unlike most mechanical devices, our limbs, bones, sinews, and muscles were not designed to repeat identical motions. Think of free-hand drawing: we draw a line, and the line we draw is the trace of the motion in space of our arm and hand. The line itself may be straight, curving, circular, or whatever, yet no matter how hard we try, we'll never manage to make another just like the one we just made. We shall, at best, make a similar one. Some awareness of this state of affairs is the basis of many social practices and cultural technologies that still pervade our daily life: personal signatures, for example, are to this day required to validate some documents and contracts, never mind that most of these documents are now digital copies, rendering the insertion of a digitized signature, mimicking an autographic one, technically absurd and legally meaningless—a paradox that even Blockchain technologies are unlikely to overcome anytime soon.

4

In the old world of pen and paper, however, autographic signatures made sense, and worked well: all signatures belonging to the same person were expected to be similar, all showing the same individual style or traits—a unique combination of ductus, stroke, and other elements that expert graphologists could analyze and validate when necessary. To the best of my knowledge, a number of central banks around the world still hold the original signature I filed when I first opened a checking account in their respective countries, now a long time ago: this archetypal autographic signature was meant to be the touchstone against which to check any further signature of mine, in case of litigation. The modern system of autographic identification assumes that all signatures by the same person will forever be recognizably similar—but never identical. From this assumption it also follows that, should two identical signatures from the same person show up, either or both may be forged.

This corollary was brought to my attention not by a scientist, but by a school headmaster, comparing two signatures of my father's on two of my school reports—the first being my father's, the second my own copy of it. Unfortunately, it was an identical copy: laboriously concocted by dint of one of the reproduction technologies then available to a child: tracing, pouncing, perforated dot joining—I forget. The point is that my copy looked like a carbon copy of the original, and no original signature ever does. That's how I learned, to my detriment, that no human hand ever repeats the exact same motions, unless by accident, or tiresome training.

Many years later I met a dentist who claimed she was the best in the world at pulling out tooth number 16—but not its symmetrical counterpart 26—because she had been doing only that since the beginning of her career, and her hand had become perfect and infallible in that gesture. The expression I remember she used, "automatism of the gesture," was very similar to that introduced early in the twentieth century by Frank Gilbreth, the Progressive Era founder of time and motion studies, whose theories influenced the rise of the modern moving assembly line—on whom more will be said in the next chapter. I have no doubt that just like the best dentists or surgeons, successful musicians and professional sportspeople (tennis players, for example) may acquire over time the capacity to repeat the same precise gestures or strokes as often as needed. But this comes at the cost of a lifetime of training, because one must teach the human body to function in a way that goes against its nature. In most other cases, the iron rule of hand-making still applies: variability and similarity are the lot of all things handmade. If you want two signatures to be identical, use a rubber stamp. But that won't be a signature anymore, because that sign will be machine-made; the rubber stamp is a copy machine, albeit a primitive one.[3]

1.2   Mechanical Machine-Making

The first, archetypal mechanical machine—the device where the technical logic of mechanical machine-making first revealed itself with immediate evidence—was, indeed, a copy machine: Johannes Gutenberg's print with movable type. Gutenberg was a goldsmith, and the idea that some graphic signs could be cast in metal and reproduced from a mold or matrix was well known within his trade; artisans of all ilk had practiced metal casting since time immemorial. When coupled with the alphabet, however, metal casting became a revolutionary media technology. Alphabets use a limited number of graphic signs (the letters of the alphabet) to notate the totality of human thought (technically, to record the totality of sounds made by the human voice); thanks to the combinatory logic of the alphabet, a small set of molds can be used to mass-produce all the movable metal type needed to print all the thoughts that humans can think, all the words that humans can say, and all the books that humans can write. Modular design by interchangeable parts was not invented by gun manufacturers Eli Whitney or Samuel Colt in the nineteenth century, as we still read in many histories of technology;[4] it was invented by whomever invented the phonetic alphabet (even though it does not appear we have a name and a date for that). As there are only so many letters in each alphabet, each letter is used many times over, in all kinds of combinations, to form words; for the same reason, when printing from movable type, the upfront cost of the mold needed to cast each letterform is amortized by repeated use. This is what today we call "economies of scale," and this is what Gutenberg must have had in mind, even if scholars at the end of the Middle Ages would not have had a name and a theory to account for that.

Economies of scale in a mechanical, matrix-based reproduction workflow are in reverse proportion to the number of variations in the set being reproduced, so when printing alphabetical text, for example, economies of scale would be maximized if the alphabet consisted of only two signs; conversely, savings would tend to zero if the number of different signs were infinite, and each sign would be reproduced from its matrix or mold only once—as if each mold were a disposable one-off. The phonetic alphabet—with its limited set of standardized graphic signs—was the software that made the hardware of print with moveable type viable.[5] In the case of notational systems consisting of millions of different signs, printing from type made from reusable molds, albeit perfectly possible, wouldn't save much time or money, and in fact similar processes were in use in a number of Eastern countries before Gutenberg started printing books that way—but in the absence of the alphabetical edge, so to speak, that technology never caught on.

In the case of early modern printing, the same technical logic we have seen at play for the mass production of movable type applied again, at a more visible scale, to the reproduction of each page printed from a mechanical matrix inserted in a printing press. Lines of text from metal type could be combined with woodcut images, albeit not with images engraved on metal plates (on which text, when needed, had to be manually incised); regardless of the components thus assembled, the printer at some point obtained a mechanical matrix ready for pressing on a paper sheet, on which the matrix would leave its imprint, or print.

When printing from a mechanical matrix, the decline in unit costs is dramatically affected by the extent of the print run: in theory (and excepting the costs of paper, ink, etc.) if the number of copies were infinite, the cost per copy would be zero. Economists today would describe that as a mathematical function, and an instance of the law of decreasing marginal costs. Continuous functions did not exist in the Renaissance, yet we may assume that the spirit of the game would not have been lost on Renaissance printers: if you try to amortize the cost of a mechanical matrix by printing more copies, the cost of the next copy you make will be less than the cost of the last copy you made; hence the incentive to keep printing, for as long as possible—regardless of the decline in the quality of the copies toward the end of each print run: *the more copies you make, the cheaper each copy will be.*

This, in a nutshell, is how we achieve economies of scale by mass-producing more identical copies of the same product. To the limit (again, a notion that would have made little sense before the invention of differential calculus), if we make just one product, just one, always the same, and we can make it equally for all, we shall have made it the cheapest. Almost six centuries after the start of the print revolution, and more than two centuries after the start of the industrial revolution, today we also know that customers who like cheaper prices may not always like identical products. This problem did not exist in a pre-modern, artisan world, where everything was handmade, hence looking different at all times, and where identical copies were a rare exception. But as soon as identical copies started to be more generally available, some started to strongly dislike them.

In the last quarter of the fifteenth century, when printed books began to compete with traditional scribal copies, many scholars were at first far from thrilled. I suspect that if the Humanists could have spoken in today's language they would have complained that books made by that barbaric German labor-saving machine looked cheap. A few centuries later John Ruskin, among others, vented the same dislike (and based his repudiation of the totality of the machine-made environment on the same visceral aversion to the making of identical copies).[6] For sure, if we compare an early printed book with a coeval illuminated manuscript, which was a unique work of art often made to order and at vast expense by a team of highly specialized scribes and miniaturists, printed books looked plain, unadorned, and cheap. But they didn't only *look* cheap. They were cheap, too. A book in print could be, it seems, up to four hundred times cheaper than a scribal copy of the same text, against which the book in print would have been competing on the same (limited) market.[7]

It is easy to see why, in spite of the Humanists' misgivings, printed books soon replaced scribal copies throughout Europe: enough customers must have concluded that, even if they did not like the look of printed books that much, they would still buy them due to overwhelming convenience and price. The transition from the old technology to the new one took, altogether, less than one generation. Electronic books—still generally seen as the natural nemesis and successor to books in print—have already been around for longer, and at the time of this writing they have not even started to effectively compete against traditional printed media. They might have, one would argue, if the price difference between a printed book and its electronic version today were similar to the price difference between a scribal book and its printed version at the end of the Middle Ages—or if prices of e-books today reflected their actual production costs (which are basically zero; indeed, this is what an e-book costs if, instead of buying it, one downloads it from a hacker's site, which gives away a stolen copy).

We now know the print revolution in early modern Europe was only the harbinger of more to come. Gutenberg standardized and mass-produced metal type and books. With the industrial revolution the technical logic of matrix-based mechanical mass production was applied to the way we make almost everything: the fewer different items we make, the more identical copies of each item we can make (hence the logic of making more different things out of fewer modular components); and the more identical copies we make, the cheaper each copy will be, thus maximizing economies of scale. Customers may not like the newly standardized or modular industrial products at the start, but they will be won over by price. The most emblematic success story in this respect is that of the automobile, not only because the automobile was one of the core technical objects of the twentieth century, but also due to the inspiration the car industry and its modes of production offered to modernist architects and designers—first and foremost, and famously, to Le Corbusier.

The story of Henry Ford's Model T is well known, and it has been told many times. Before the Model T, cars were made to order. Customers went to see an artisan car maker and they ordered a car to be made to their taste, much as they would have ordered a custom-made shirt, or suit, from a tailor. Then Henry Ford decided that, instead of waiting for each customer to tell him the car he wanted and making it to specs, he would prefabricate and offer to each customer the same ready-made car. By making more identical copies of the same item (a mass-produced, standardized car) he would be able to sell each car for less. As customers were used to choosing some core features of the cars they ordered, Ford shrewdly advertized as a "modern" novelty what was in fact a crucial limitation of his offer— hence the famous tagline "you can still choose the color so long as it is black." But the decisive argument to persuade Ford's customers to buy a car that was not exactly the car they wanted was one that needed no advertizing: Ford's Model T cost significantly less than any similar car made to order in the traditional, artisanal way.[8]

The way mass production and economies of scale played out in the fabrication of automobiles was not conceptually different from what we have seen in the case of Gutenberg's print: most metal parts to be assembled in a car were either metal sheets printed by mechanical presses or metal parts molten and cast in molds: casts, stamps, dies, or molds are all mechanical matrixes, the cost of which is amortized by repeated use. But while Gutenberg's typesetting (the assembly of mass-produced type into lines of text) was entirely manual, Henry Ford also started to envisage ways to mechanize, or automate, the assemblage process itself. That resulted in the introduction of the moving assembly line, which came into service a few years after the start of production of the Model T, but was determinant in abating its production costs. Workers in a moving assembly line repeat a very limited sequence of simple motions many times over, following the assumption, then frequently held, that this would make them faster and more efficient at their task. As we shall see in the next chapter the economics of the assembly line, driven by Frank Gilbreth's time and motion studies (1909) and by Frederick Taylor's theories on task management and subdivision of labor (1911), while similar in spirit, so to speak, must be seen as conceptually distinct from the economics of matrix-based, mechanical mass production: the latter standardized the product through the reproduction of identical parts; the former standardized the production process through the repetition of identical motions.

All this happened in America in the years immediately before World War I. Seen from Europe a few years later, this was not only a resounding industrial success story. For young Le Corbusier, Fordism was a beacon of hope: it showed a pathway to reform architecture and—Corb was never a man of limited ambition—the world. Corb's argument, as we can elicit it from his writings of the early 1920s, followed from a simple parallel. The artisanal automobile was custom-made, and expensive. Ford brought to market a mass-produced, industrial product: the same car for all. The economies of scale generated by mass production made Ford's standardized car affordable: what used to be an artisanal product for the few became an industrial product for the many. Architecture, young Corb mused, is expensive because each house is custom-made, made to order, and made by hand. Why should we not do with architecture what Henry Ford did with the automobile? Standardized, industrially mass-produced houses will be cheaper than artisanally made ones. We could then offer a new house to all that need one. In the pursuit of economies of scale, all these machine-made houses will be the same, and will look the same. Corb also thought that, just to be on the safe side, he should design them all.[9]

From the start, Corb had a clear notion of what these mass-produced houses would look like. He first published his project for a "serial house Citrohan—not to say Citroën" in 1921, in the thirteenth issue of his magazine *L'Esprit Nouveau*, then republished it in his book *Vers une architecture* (1923).[10] André Citroën was known for his innovative engineering, and he was the first car maker in France to adopt Ford-style assembly lines. Corb even tried to entice Citroën himself to fund his Pavillon de l'Esprit Nouveau at the 1925 Arts Décoratifs exhibition;[11] Citroën declined, and Corb got a check from the Voisin car company (which went into bankruptcy soon thereafter). The famous Voisin Plan than ensued clearly illustrated Corb's vision for a car-driven future: new cities made to measure for automobiles, with new buildings built as if they were automobiles.[12] The first part of the plan was easy—the Italians had already started building separate highways for speeding cars (which in fact Corb called *autostrades*, using the Italian neologism). But the second part of the plan was not easy at all, because back then—as now, in fact—nobody really knew how to make a house in a factory.

Although Corb didn't elaborate, his Citrohan House of 1921 looks as if it were made of reinforced concrete. But reinforced concrete (let's not forget, in 1921 still a new and relatively untested technology) is better cast on site—a fairly artisanal and labor-intensive operation—than prefabricated off-site and then assembled; after many, mostly unfortunate, experiments with the industrial prefabrication of concrete components in the 1950s and 1960s, this is still the consensus of the technical professions (at least as far as housing is concerned). Only a few years later Corb concluded that while load-bearing concrete structures had to be made on site, the non-structural facade walls, mostly consisting of metal and glass, could be made in factories, then shipped and pieced together on the building itself. Metal frames and glass plates: this is indeed what factories of the twentieth century were good at mass-producing, and the glass and metal curtain wall, which Corb introduced in his buildings of the late 1920s and Mies famously perfected in his American   11 high rises after World War II, became the outward and visible sign of the modern, industrial way of making: industrial mass production generates economies of scale that are proportional to the number of identical copies we make; identicality is both the index and the symbol of the Taylorized, machine-made environment.[13] This is why all the windows of the Seagram Building were designed to look the same from all sides and at all times of day and night, even when the interior blinds are pulled; this is what industrial economies of scale were always meant to look like, never mind that the bronze-hued metal frames of the Seagram's curtain wall made the building one of the most expensive of its time.

1.3   Digital Making

It is important to retain the technical logic of the matrix-based, mechanical-industrial way of making—as the technical logic of digital making is its opposite. This is because, in most cases, digital making is not matrix-based—a banal technical truism with enormous consequences. Example: let's compare any vintage contact-based, mechanical printing machine with current digital printing technologies. I do not know how many among the readers will have ever seen a mimeograph such as the one I still used, mostly for extra-curricular purposes, when I was in high school in the 1970s (the one we used was actually a British-made Gestetner Cyclostyle; the Gestetner company is better known among architects for having commissioned Highpoint I, a modernist apartment block built in 1935 in London by Berthold Lubetkin). A matrix-based mechanical duplicator, the mimeograph became an iconic communication technology in the 1960s and 1970s—used by hippies, artists, political activists, trade unions, churches, and revolutionaries of all sorts. The matrix from which it printed was in fact a flexible stencil—in my days, made of some kind of soft, ductile plastic: the stencil was engraved by typing on it (with the typewriter in "stencil" mode), or it was manually incised with the help of special styluses; the stencil was then hooked to a rotary drum operated with a manual crank; each rotation of the drum inked the stencil grooves and drew in a blank sheet of paper onto which the stencil left an imprint by direct contact. The stencil itself, which was specific to each machine and had to be bought in specialized shops, was not, as far as I remember, particularly expensive, but its preparation was laborious and time consuming; manual engraving in particular was a delicate operation, as the engraver had to be careful not to pierce the plastic coating. In a normal commercial environment, all this would have translated into costs. Even operating at zero labor cost, as we did, we had to be careful to make the valuable stencil last for as long as possible, by making as many imprints of it as it could physically endure. Sometimes the stencil would give way little by little, and its final collapse would be announced by the worsening quality of the copies; sometimes the stencil tore apart without warning, sending ink in all directions— generally hitting me first, right in the face. Regardless of how and when the print run would end, the iron law of decreasing marginal costs determined, more or less unconsciously or subliminally, the viability and outreach of all our printing projects. We knew nothing of that law, yet we lived by it.

Now let's compare all that with what happens when we use any of today's personal digital printers, either laser or inkjet. First of all, for the kind of projects I was referring to, and for which in 1976 mimeographed printouts were the only mode of dissemination, today we may not need any printing at all. What back then we could only print on paper (or scream into a megaphone) today we would tweet or post or blog. But that's another story. Let's assume for the sake of this argument that there may still be something we really need to print, for whatever reason. Laser printers and ink-jet printers do not contain any mechanical matrix or stencil. Desktop laser printers in particular do not look very different, on the outside, from old mimeographs, and they do contain a rotating drum that prints by direct contact (the laser beam first writes on the drum; the drum then attracts ink powder and transfers images onto paper); but printing is made line by line, as the drum sequentially impresses horizontal strips of dots, called raster lines or scan lines. The matrix-less technical logic of digital printing is more transparent if we look at an inkjet printer at work: the print head moves horizontally across the page as the page advances through the printer, and the nozzles in the print head depose drops of ink where needed. Each page is notionally divided into a regular grid, or raster, of pixels or dots; the memory (or bitmap) of these dots, each identified by its coordinates on the page, then automatically translated into code to drive the motions of the print head, is in a sense the matrix, or memory, of each 13 page being printed; but this matrix is made of bits and bytes—i.e., it is pure information, with no physical weight, and it is incessantly refreshed (canceled and rewritten) at almost no cost (or at digital processing costs, which are, again, almost zero).

This means that after a page is printed, the next to be printed will be, for all practical purposes, an entirely new page, as no material leftover from the printing of the first page will be reused for the printing of the second page—even if the second is identical to the first. Consequently, as the printing of each page must be seen, conceptually, as a stand-alone, one-off operation, regardless of precedent and sequel, the unit cost for the printing of each page in any given print run will always be the same. A digital printer will print one hundred or one thousand or one million identical copies of the same page at the same cost per page; it will print one hundred copies of the same page, or one hundred different pages, at the same cost. As a result, when printing digitally there won't be any need to ever "scale up" production in the pursuit of economies of scale, and as the cost of each copy is always the same, regardless of the number of copies made, the curve of marginal costs isn't a curve at all in this instance—it is a horizontal line (which is what economists call a flat marginal cost).[14]

This logic is even more evident if, rather than text and images on paper, we print three-dimensional objects. Imagine we are in the business of making statuettes, and we 3D print these statuettes using a small desktop 3D printer. After we have printed a statuette of, say, a cat, and we have calculated that the production cost of that statuette is, say, one dollar, we may imagine that if we 3D print one thousand identical copies of that statuette the unit cost of each statuette will decline, and will keep declining the more identical statuettes we make. This indeed would be the case if we used a mold to cast that statuette in metal, or plastic. In that case, we would use the same mold many times over, thus spreading its upfront cost over many statuettes, which will make each statuette cheaper. But there is no mold in that 3D printer, because each statuette is made as if out of thin air, by deposing layers of material squeezed or sprayed out of a nozzle; in that sense, each statuette we make is a new one, even if all statuettes in that series are identical; the printing of each of those cat statuettes will always take the same time and, when all cost factors are taken into account, cost the same, no matter how many identical statuettes we make. We can 3D print one thousand identical cats, and one thousand different cats, at the same cost per cat (barring the "authorial" cost of designing each different cat, evidently).

This way of digital making has a name—it is called digital mass customization. The term, which existed before the rise of digital making with a different meaning,[15] results from the conflation of two different and apparently incompatible terms: mass, as in mass production (that's what factories do—that's the mechanical-industrial way of making); and customization, which is what artisans do, making something to a customer's order. Each custom-made item can be different from all others—it doesn't necessarily have to be, but that's irrelevant from the point of view of an artisan, because the ideal artisan works on demand, and does not prefabricate. In theory, artisans cannot mass-produce, and industrial-mechanical factories cannot customize. Digital making should now allow us to mass-produce *and* customize, thus joining the advantages of artisan customization to the costs of machinic production, already way cheaper than artisan labor—at least, in all advanced economies, and increasingly elsewhere.[16] Hand-made pieces are always different (variability); matrix-based, mechanical machine-made pieces are always the same (identicality); digitally made pieces can be all different or all the same, as need be, at the same unit cost (digital variability, or differentiality, which defines non-standard serial production, or digital mass customization).[17]

Designers have been discussing digital mass customization as well as its design corollary, digital craft, for some time now. The idea emerged in the 1990s and it has been one of the staples of digital design theory since. More recently, some economists and technologists have also started to take notice.[18] And rightly so, as digital mass customization is one of the most disruptive ideas of our times, as important today as the idea of mass production was one hundred years ago. Digital mass customization will mark post-industrial societies in the same way as mass production marked the industrial revolution. In the new digital world, making more of the same will not make anything cheaper. Hence the question: what happens if we shift from a society, a culture, and a civilization, dominated by a relentless and ubiquitous quest for scale, to a new techno-social environment where scale does not matter? What happens when *economies of scale* are replaced by a *new economy without scale*? The novelty is so colossal that we should not blame politicians (but also industrialists, philosophers, sociologists, and many academics and opinion makers) for having so far been incapable of grasping the enormity of the interests at stake. For, born and bred in a mechanical-industrial world, we are all so used to the modernist logic of scale that we may not even conceive of a technical alternative to it—let alone imagine a world, and a view of the world, where economies of scale have ceased to be.

1.4 Beyond the Anthropocene: A New Economy without Scale

A few centuries of industrial-mechanical making have made us used to thinking that if we make more copies of any given item, we can make them for less. With some logic, this rule has translated into a similar one, which relates not to costs but to prices—not to the logic of production but to the logic of consumption. We live in a world where pretty much everyone is intimately and often irrationally persuaded that if one buys more of any given item, one should pay less. Example: let's assume that we want to buy a new car (and for the sake of the argument we shall also assume that someone may still want to actually buy a new car rather than lease it). Even in that case, these days we wouldn't necessarily haggle at length on the car's listed price. But if you run a small company, and you buy five identical cars on the same day, you would expect to get an almost automatic discount on the listed price of each; if you run a big company, and you buy one hundred of those cars in bulk, you will likely get a big rebate. There are many, mostly psychological, explanations for this phenomenon—such as, the customer may wield a bigger bargaining power; or the salesperson, often paid on a commission basis, will be eager to sign the deal of her lifetime, and so on. But all these cars, regardless of how many you buy, are likely to have already left the factory at the time of your purchase, and no sale deal will retroactively change their production costs. Even more bizarrely, why do we apply this same logic of scale when we buy non-industrial products, such as, for example, potatoes? If we buy ten pounds of potatoes, we expect a price per pound which is less than we pay if we buy just one pound; if we buy potatoes in bulk, we will expect and get a significant reduction on their retail price. But potatoes are not factory made; there are no economies of scale in the way tubers and roots grow in the furrows of the earth (even if harvesting will be more efficient, I guess, in big fields than in small gardens). And then, of course, we think in the same terms even when we buy non-reproducible, custom-made services; so we would typically expect to get a rebate on the posted daily rate of a hotel room if we book it for seven consecutive days, and an even bigger rebate if we stay for one month.

In many similar cases, bigger transactions do sometimes entail a reduction in unit costs due to a number of organizational factors. Each transaction has fixed administrative costs, which are often the same regardless of size: for example, the paperwork needed to import a pair of shoes into the US is the same if you bring in your own pair of sneakers to go for a walk, or one million pairs of identical sneakers for global distribution. Therefore, it often makes economic sense to aggregate supply and demand, spreading the same fixed costs over fewer, bigger transactions, hence reducing the part of fixed costs to be absorbed by each individual item. These would be actual economies of scale—in this instance applying to commerce and services, not to manufacturing. But this does not and cannot by itself account for the deep-seated, engrained mentality and general expectations of millions of customers around the world, all intimately persuaded that—no matter what you buy—*if you buy more, you should pay less*. The permanent quest for scale brought about by the technical logic of mechanical machine-making has turned into a way of thinking, detached from any technical justification it may once have had—into an ideology that has shaped the world we live in and still pervades many aspects of our daily life.

Imagine now that you are an archetypical modern industrialist—and an exemplar and champion of the industrial-mechanical worldview: you know by heart all the written and unwritten rules and principles of mechanical machine-making, and you want to follow that playbook to grow your business. Let your business be making teapots this time—not cat statuettes. You know that to maximize economies of scale, and to make more copies of fewer items, you should first of all aim at the design of a generic, universal teapot—a global standard that can ideally sell to all in all places. Then you should build your factory in the most convenient location, based on production costs and related factors. Let's assume that this 17 location turns out to be in central Mongolia. You will build your factory there, and you will try to make it as big as possible, in order to make as many identical teapots as you can sell. To sell those teapots, however, you will need not only transportation, but also unimpeded access to markets—to as many markets as you can reach, and as vast as possible. In short: to make cheaper teapots, you will need a bigger factory, so you can make more identical teapots. But to sell more teapots, you will need a bigger market. Hence another golden rule of the industrial-mechanical playbook: bigger markets mean cheaper goods (for consumers; and/or bigger profits for producers). *Bigger is Cheaper*.

The size of your primary market is generally determined by the borders of the country you're in. Some foreign countries may not allow the import of your teapot. Others will oblige you to apply for permissions and pay custom duties. The bottom line: to maximize economies of scale in the production of your teapot, you will need first and foremost to expand your home market, meaning that at some point, after each resident of your country has bought your teapot, you will need your country to get bigger, and/or the size of its teapot-buying population to grow. There are not many ways to achieve that. The first idea that comes to mind is to have the army of your country invade a neighboring country. In the course of the nineteenth century European capitalists concluded that a custom union (such as the German Zollverein, started in 1833) could in some cases be more expedient than an all-out military invasion. Other means to the same end: colonialism, common markets, annexations, national unifications, unimpeded immigration, mergers and acquisitions (not of companies but of countries and territories), and so on. Regardless of many still touted Romantic ideals of nationhood and patriotism, the rise of nation states in the nineteenth century was often determined, to some extent, by a technical quest for scale—or, to be precise: by the need to arrive at a national market in a size that would optimize the economies of scale achievable in each phase of the process of industrialization. This ideal size changes over time, following the changes in the technical factors on which it is based. Consequently, as digital production doesn't depend on scale, national markets—as well as the nation states that created, nurtured, and protected national markets during the industrial revolution—
18   are now about to lose all (or most) of their original reasons to exist,

and they are doomed to extinction. This is not the least of the many consequences of our current transition from an old world dominated by economies of scale to a new economy without scale.

Other instances of the same general trend are more immediately tangible. As mentioned above, it was and still is customary in many trades to aggregate some transactions in order to amortize some of their fixed costs. But what if the paperwork (literal, or metaphorical) responsible for those fixed costs could be automated, and carried out by some electronic algorithms in almost no time, and at almost no cost? When the paperwork for a car rental took one hour of both the customer's and the rental agent's time, nobody rented a car for less than a day. Today, we can rent a car, find it, and unlock it with a tap on an app. The time, fixed costs, and administrative overhead needed to process such electronic rental agreements are negligible; with some logic, these rentals can have durations as short as thirty minutes, and the cost per hour of the rental does not get cheaper if the duration of the rental gets longer. No discount for buying more: that's one tenet of the modern industrial-mechanical world that has already been jettisoned. Why? Because the fixed costs that would have been amortized by a bigger purchase are now almost equal to zero from the start.

Until recently, nobody would have dreamt of signing a legally binding agreement to rent a bike for fifteen minutes, and for sure nobody would have done so to monetize a transaction of which the total commercial value is almost nothing. The rent-a-bike schemes now frequent in cities around the world do just that. The granularity and versatility of these online-only, algorithm-driven micro-contracts have already brought to market transactions so small and so idiosyncratic that until recently could only have existed in an informal, gift-based economy ("Can I have your bike? I'll be right back"; "Can I have your apartment this weekend? I'll leave a bottle of champagne in your refrigerator"). Many more examples could follow: for better or worse, the scaleless nature of the digital way of making—inherent in the technical logic of all things digital—has already started to change the way we manage many practical aspects of our daily lives, even if it does not seem so far to have significantly changed our way of thinking.[19]

One can easily imagine a believable computational alternative to the global teapot megafactory I just caricatured. Teapots can be easily 3D printed. Not everyone may have a suitable 3D printer at home, but a local microfactory, or Fablab-like installation, as imagined by Neil Gershenfeld almost twenty years ago,[20] could provide that. The file with the design of the teapot could be downloaded and tweaked (or customized) as needed; the clay to make the teapot may be locally available, or would not come from afar; the electricity to power the printer and the oven to bake the clay would be locally generated from renewable sources.

I do not think anyone can provide reliable econometric studies to compare facts and figures on this matter, but I would argue that, by and large, this 3D printed teapot, made where you need it, when you need it, as you need it—on site, on specs, and on demand—shouldn't be more expensive than a globally manufactured one, even without taking into account the transport of global products from their faraway sites of production. The metrics of this comparison would change drastically, however, if, alongside the direct costs of transportation, we take into account their environmental costs. Again, in the absence of facts and figures, which would be difficult to calculate anyway, we are left at the mercy of competing ideological interpretations. Many have rightly pointed out that computers and electronic communications in general use a lot of electricity and produce plenty of material waste.[21] No one will object to that. Common sense would still however suggest (pending suitable metrics and verification) that downloading a few lines of code to print a teapot in a Fablab at the corner of the street uses less energy, and has a smaller carbon footprint, than shipping the same teapot, or a very similar one, from the global factory where it is mass-produced (in my fictional case, central Mongolia) to where you need it. One may or may not want to take environmental costs, and carbon footprint, into account. The vast majority of the design community decided long ago that climate change and global warming matter—and this was before the global pandemic that started in early 2020 added new socio-technical considerations to the urgency of our environmental challenges.

1.5   The Collapse of the Modern Way of Making

Digital designers and critics (myself included) have long claimed that the electronic transmission of information is cheaper, faster, smarter, more efficient, and more environmentally responsible that the mechanical transportation of people and goods; and that digital mass customization in general is cheaper, faster, smarter, more efficient, and more environmentally responsible than mechanical mass production. We have been saying that since the mid-1990s, and mostly to no avail.[22] Many arguments for and against digital mass customization have already been discussed at length. One of them is, again, a matter of ideological choice: we need all things to be made differently because we are all different; we need all things to be made the same because we are all equal. In purely technical terms, there are only a few instances where we can make a clear case for the superiority of digital mass customization: think of tooth implants, or hip replacements. These medical appliances cannot be standardized (a "same tooth for all" policy does not appear to have been tried anywhere to date), and they are now better and cheaper when digitally mass-customized than custom-made in the traditional, artisanal way. At the opposite end of the desirability of digital mass customization, my BIC ballpoint pen does not need to be any different from anyone else's. All other cases of digital mass customization fall, with nuances, in a gray area between these two extremes; generally speaking, digital mass customization would appear to suit some products more than others. Yet the wastefulness of mechanical mass production, when compared to the inherent frugality of digital mass customization, has been proved many times over. This argument may seem counterintuitive to many used to lambasting the showy profligacy, lavishness, or purely cosmetic, narcissistic customization of many contemporary products of digital design. But this is not what is at stake here; digital mass customization is a technical logic, not an architectural style.[23]

Architectural history offers one of the most transparent illustrations of the trade-offs and costs inherent in the pursuit of the modernist logic of mass production. While reinforced concrete is best cast on site, structural components in steel and in prestressed reinforced concrete are more often prefabricated off-site, then transported and assembled on site. Mass-produced in factories in compliance with strict industrial standards and quality controls, steel I-beams, for example, were for long seen as a typical—indeed a quintessential—industrial product. Things are different now, as digital mass customization has started to work its way into steel mills as it has elsewhere. But this was the logic that Corb, Mies, or Gropius would have been familiar with in the 1930s or in the 1960s, when steel I-beams were emblems of industrial modernity, and they visually symbolized, with almost brutal self-evidence, the technical logic of mass production that determined the relation between their design and their cost: making more identical beams will make each of them cheaper.[24]

Mies famously translated this technical logic into an art form. The mullions in the curtain walls of his high rises are not structural, but the roof of his New National Gallery in Berlin, for example, is a grid of identical I-beams that crisscross each other, creating ideally an isotropic plate, even though nobody knows which beams are continuous and which are cut at each intersection: nobody can tell because the beams are, or at least look, as you would expect—Mies being Mies—all the same.[25] If you are an engineer, and you choose to use the same I-beams all over your structure, you will calculate the size of the section under maximum stress, and then let all other sections be the same. That way, you will use way more steel than you would if you made each section as big as needed. In compensation, you will be able to buy a cheaper batch of identical I-beams—never mind that all of them, but one, will be oversize.

However, let's do something now that many twentieth-century engineers never had to do. Let's compare the savings thus gained from industrial mass production with the costs of the raw materials that go wasted due to the adoption of standardized structural components. Mid-twentieth-century designers often didn't have to do that calculus because for them the cost of raw materials, as the cost of energy, was supposed to be more or less irrelevant. In their cost equation those factors had a coefficient of almost zero. To them, only the cost of labor mattered, and to save on labor costs the industrial-mechanical technical logic offered only one, famously effective, tried-and-tested solution: standardization and mass production. And by the way, a somewhat similar logic applies, with similar architectural results, if we build in reinforced concrete and most labor is performed on site: a generic, simplified structure will save on labor but at the cost of oversizing; a more meticulously designed structure (with variable sections along each beam, for example) may save on materials but add to labor costs; since the invention of reinforced concrete designers have chosen either strategy based on the different incidence of labor cost and of the cost of raw materials in different markets and over time.

Today energy costs and the costs of raw materials are far from irrelevant, and we can expect that the cost of non-renewable resources in particular will keep growing—even more so when environmental factors are taken into account. The cost of labor, in turn, will be increasingly affected by robotic automation—this is already the case in many factory-based industries, and soon the same will happen in the building and construction industry. While the cost of manual labor may still rise, the cost of robotic labor is most likely going to decline.[26] So our cost equation has changed—it is, in a sense, the exact opposite of what it was for our modernist predecessors. Some have suggested, perhaps optimistically, that robotic labor and renewable energies may soon be generally available at zero costs or almost. This may or may not happen but regardless, we know as a fact that using today's digital technologies we can already mass-produce variations at no extra costs: in the building and construction industry in particular, we may therefore more realistically be heading for production *without economies of scale* than for production *without costs* (hence heading for a flat marginal costs economy, not for a zero marginal costs society).[27] In an ideal, fully digitized design-to-production workflow, making a given number of structural I-beams all the same or all different, as any savvy engineer would like them to be, should in theory cost the same. When that happens, engineers can go back to the drawing board, knowing that elegant, thrifty, frugal structural design is not going to cost more than the stodgy, brutal, and just plain stupid structures that marked late twentieth-century civil engineering—structures determined not by the natural logic of the materials we use, but by the industrial logic of the factories those materials come from.[28]

So it will be seen that the industrial-mechanical logic of mass production, standardization, and economies of scale, albeit eminently successful throughout the twentieth century, came at huge environmental and social costs: it forced workers to do jobs ("alienated" jobs, as someone said)[29] that they often did not like; it forced consumers to buy standardized products that often did not fully match their needs; it entailed wasteful environmental costs that we now know could not have been sustained for long. This is why, with the rising awareness of global warming and climate change this

industrial-mechanical model, which many today call "anthropocene," came to be seen as "unsustainable." Digital designers, without
using either term, have been vocally advocating a mass-customized,
information-driven, computational alternative to the mechanical-
industrial model since the early 1990s. They did not have to claim
that the old model was "unsustainable," because their claim was
bolder: as they argued, the old model is simply obsolete—deemed
to extinction and due to be replaced by another one which, unlike
the old one, offers some hope for our future. Before the COVID-19
pandemic the notion of the "unsustainability" of the modern industrial model was often a fig leaf, used by all and sundry to plead for
everything and its opposite. Today, the term itself is simply plain
wrong. It is wrong because the industrial-mechanical way of making
is no longer "unsustainable"—it is already, literally, *unsustained*,
having already collapsed in full: precipitously, catastrophically, suddenly, and sooner than we all expected; eliminated not by global
warming but by a global pandemic; not by climate change, but by
viral change.

1.6   The Teachings of the Pandemic

Who doesn't remember the dot-com crash?[30] It began in March 2000,
twenty years almost to the day before the start of the COVID-19 pandemic in Europe and in the US. But for younger readers, who may
indeed not remember, here comes a brief recap. Around the mid-
1990s many started to claim that a major technological revolution
was underway, and that digital technologies were about to change
the world—and to change it for the better. Architects and designers
in particular were enthralled by the creative potentials of the new
digital tools for design and fabrication; digital mass customization,
then a new idea, promised a complete reversal of the technical logic
of industrial modernity. At the same time sociologists and town
planners were trying to make sense of a new information technology with the potential to upend all known patterns of use of urban
space, and of cities in general: the internet was still a relatively new
concept (many still called it "the information superhighway" or the
"infobahn"), yet some started to point out that, with the rise of
the internet, many human activities were inevitably poised to
migrate from physical space to what was then called "cyberspace"

(i.e., again, the internet): Amazon sold its first book in the spring of 1995. The company was then called, as it still is, Amazon.com.

In the years that followed every company with a dot and a "com" in its name, as per its URL, or internet address, seemed destined for the brightest future. So was the internet in general, and with it, many then thought, the world economy. As the late William Mitchell pointed out in his seminal *City of Bits* (1995),[31] many things we used to do face to face can now be more easily and more efficiently done electronically: think of e-commerce, e-learning, e-working (or remote working, or telecommuting), and so on. As one proverb frequently cited at the time went: for every megabyte of memory you install on your hard disk, one square foot of retail space downtown will disappear.[32]

Strange as it may seem today, in the second half of the 1990s everyone thought that was a splendid idea. The valuation of all dot-com companies (companies doing business on the internet, or just saying they would do so at some point) soared. Between January 1995 and March 2000 the NASDAQ composite index, where many of these young companies were quoted, rose by almost 600 percent. As the then chairman of the US Federal Reserve, Alan Greenspan, famously said, "irrational exuberance" was not the only reason of that market optimism: valuations were rising because the internet made our work in general more productive, and many things easier to find, buy, or sell—hence cheaper. Thanks to the internet, we were told, we were all doing more with less: more work, more reading, more teaching, more learning, more researching, more interacting, more dating—you name it. The electronic transmission of data costs so much less than mechanical transportation of persons and goods: think of the advantage of reading a scholarly article from your office—or from your couch!—without having to travel to a faraway library. Around that time some also started to point out that the elimination of the mechanical transportation of persons and goods could be environmentally friendly (or, as we would say today, would reduce our "carbon footprint").

That seemed too good to be true—until it wasn't. The NASDAQ peaked on March 10, 2000. It lost 80 percent of its value in the eighteen months that followed. That was the dot-com crash, a.k.a. the burst of the internet bubble. Many tech companies disappeared; Amazon, for example, barely survived, after losing 88 percent of its 25 market capitalization. The NASDAQ itself took fifteen years to crawl back to its value of March 2000. In the contrite climate of those post-crash years (which were also the post-9/11 years) few still saw the internet as a benevolent, or even promising, technology. The anti-internet backlash was swift, and predictable. As many had warned from the start, technology should not replace human contact; there can be no community without physical proximity. For Christian phenomenologists, always overrepresented in the design professions, the elision of human dialogue started with the invention of alphabetic writing: if we write, we use a technology to transmit our voice in the absence of our body. For those sharing this worldview, disembodiment is the original sin of all media technologies: after that first and ancestral lapse into the abyss of mediated communication, things could only go from bad to worse; the internet was just more of the same. A few years into the new millennium the so-called social media reinvented the internet; in recent times we have learned to fear their intrusion on our privacy. Furthermore, by abolishing all traditional tools for thoughtful moderation, and giving unmediated voice to so many dissenters, outliers, and misfits, the internet is seen by many as the primary technical cause of the rise of populism. That may as well be true, regrettably, although I suspect that if I had been a Roman Catholic cleric around 1540 I would have said the same of the use of the new barbaric technology of print by the likes of John Calvin or Martin Luther.

Fast forward to the early summer of 2021. The many lockdowns we have endured in Europe since March 2020 have given me ample time to contemplate the unfolding of an unspeakable man-made catastrophe, compounded by political cynicism, criminal miscalculations, and incompetence. Throughout all this, the internet was, literally, my lifeline. For long weeks and months it was all I had. I used it to shop for groceries and other essentials, to work, to teach, to read the papers, to pay my taxes, to watch the news and occasionally a movie, to listen to the radio, to communicate with family and friends; at some point I even considered restarting my Facebook account, which I jettisoned some twelve years ago (and my reasons for doing so back then are likely still posted on my Facebook page, since forsaken).

While self-isolating in my London apartment during the grimmest, darkest days of the first British lockdown, I could see out of my living room windows the cluster of corporate skyscrapers rising from the so-called "square mile" of the City of London—glittering in the sun, when there is any, and shining at night, all lights on. They were empty, evidently, except for security and essential maintenance crew, and they remained so for months. Yet the London Stock Exchange and the global insurance and financial markets for which the City of London is an essential hub never shut down. Retail banking kept functioning too, yet I could not have visited a single branch of my bank for many months, as they all shut down when the emergency started. Some local branches—as well as most financial offices in the City—reopened with a skeleton staff in the summer of 2020, then shut down in the fall, then reopened again, following the ebb and flow of the pandemic. Reopen, sure, we are all for that—but for whom? To do what? I guess that scores of cost-cutting strategists at most banks are asking themselves that very question right now. Various successive lockdowns since the start of the pandemic have by now eloquently proven that most businesses can carry on just fine with only a fraction of the physical infrastructure that only a few months ago was considered indispensable to their trade.

As an aside, having myself spent most of my pandemic time teaching on Zoom, I can attest that most students kept studying, most lectures were taught, most papers were written, and most exams were passed while our university campuses were entirely, or almost entirely, shut down (often for eighteen months on end). We now have evidence that some online teaching can work—and pretty well, too; and that remote teaching, as well as remote working, have some advantages, and even more so when the carbon footprints and costs, including human costs, of commuting, travel, and transportation are taken into account. Remote teaching, however, just like teleworking, is hardly a recent invention: Mark Taylor, a professor of philosophy and religion now at Columbia University, started teaching online classes at Williams College in the early 1990s, and his first experiments are reported and discussed in a book he published in 1994.[33] Yet, in spite of this early start, e-teaching never really caught on. Even in more recent times the rise of Massive Online Open Courses (MOOCs) has been widely dismissed, disparaged by the global academic establishment and seen with suspicion by community activists. As a result, we all had to learn the art and science of e-teaching in a rush—the space of a weekend—in March 2020.

For some inexplicable reason—and I am still waiting for someone to tell me the rationale behind such a hare-brained decision, which must have been made soon after World War II—the east-west alignment of both runways of Heathrow Airport, fifteen miles west of London, almost coincides with the east-west axis of Oxford Street, the topographical spine of central London, so that in normal times any view of the sky over London offers the sight and sound of an uninterrupted flow of evenly spaced airliners gliding to land. For months in the spring of 2020, as travel came to a standstill, I was surprised to see still a handful in the air—and I wondered for whom, and what for. I found out. Those were passenger planes alright, but not passenger flights. The entire cabins of those planes, including their first-class suites, were being used to transport personal protective equipment and other medical appliances, imported at the last minute and at huge costs from the few countries still willing to sell any. Throughout the summer of 2020 the roar and whines and whistles of low flying airplanes in the skies over central London— usually a continuous and often maddening nuisance—remained a rarity. Only a few months before the start of the pandemic the climate activist Greta Thunberg still incited us, by words and deeds, to "flight shaming"; she can rest now—she won her battle in a big way, albeit not in any way she would have chosen. Some have even suggested that as the carbon-heavy economy of the industrial age (or anthropocene) stopped in full for some months (and kept operating below capacity in most countries throughout 2021), we may have already delayed the timeline of global warming.[34] Right to the eve of the pandemic some climate activists—as well as several sorts of self-styled "collapsologists"—were more or less openly advocating the elimination of part of the human population as the only way to save the planet: well, again, there you go.[35]

Yet, while offices, factories, schools, stores, and global transport all suddenly shut down, some commercial life—and even life in general, for those who were not infected—carried on, because farming, local artisan production, food distribution, utilities, telecommunications, and, crucially, the internet kept functioning. Many, like me, had long argued that the technical logic of mechanical modernity was obsolete, due soon to be replaced by the technical logic of electronic computation. A few weeks into the pandemic we felt, sadly, vindicated: the entire industrial-mechanical world—our enemy, in a sense—was in full meltdown and had just ceased to be, as we always said it would—although nobody could ever have imagined that the fall would come so soon, and would be so precipitous. And while the mechanical world was imploding, farmers kept farming, bakers kept baking, and thanks to the internet many of us kept working, studying, and communicating, somehow, throughout the lockdowns. The farmer, the artisan, and the internet: craft and computation—all that existed before the industrial revolution, and all that came after the mechanical age—that's what kept us going during the crisis. At the same time, the office, the factory, and the airport—the very backbone of the modern industrial-mechanical world—more or less shut down. We always claimed that digital technologies would be our future. For most of 2020 they were our present. I wish we should not have needed a global pandemic to prove our point.

The global pandemic did not only prove that a viable alternative already exists to the modern way of making.[36] It also proved that the technical logic of industrial modernity was always inherently doomed—and bound to fail. For years many have argued that internet viruses would at some point destroy the internet, and all activities depending on it (i.e., today, almost everything). That may still happen. But the coronavirus was a real virus, not a metaphorical one. It did not travel on the World Wide Web, nor did it affect the internet; it traveled by plane, boat, and rail. It was born and bred as a pure product of the industrial-mechanical age. It was a quintessential disease of the anthropocene; in a sense, a late anthropocenic disease, just like tuberculosis was an early anthropocenic one. If, when this all started, we had already been using more internet and flying less (as we are still doing now by necessity, not by choice), many lives would have been saved, because the virus would have had fewer conduits for spreading. So perhaps, in retrospect, this is exactly what we should have been doing all along.

Schools, stores, and factories have now reopened, somehow—but many offices have not, for example, and many of us are still working from home, and often shopping from home. This is because the pandemic has shown that the modern way of working—the mechanical, "anthropocenic" way of working—is no longer our only option. We now have evidence that in many cases viable electronic alternatives to the mechanical transportation of persons and goods do exist and, when used with due precautions and within reasonable limits, can work pretty well. Remote working can already effectively replace plenty of facetime, thus making plenty of human travel unnecessary. The alternative to air travel is not sailing boat travel; it's Zoom.

Service work and blue-collar work cannot yet be despatialized as effectively as white-collar work, but that's not too far away in the future either: automated logistics, fulfilment, and fully automated robotic fabrication are already current in some industries. As seen earlier in this chapter robotic factories, and digital making in general, are mostly immune to economies of scale; as they do not need to "scale up" to break even, they can be located closer to their markets, thus reducing the global transportation of mass-produced goods and components. Besides, robotic work was not disrupted by the contagion; digital artisans, working as all artisans in small shops and often in relative isolation, were only marginally affected by the lockdowns, and many kept working and producing during the pandemic. Anecdotally, but meaningfully, I know that some among my friends and colleagues, like Manuel Jimenez Garcia at the Bartlett, or Jenny Sabin at Cornell, at the very start of the pandemic converted their 3D printers and robotic arms to produce protective equipment for medics and hospital workers—on site, on specs, and on demand. Because this is indeed the point—this is what robotic fabrication was always meant to do: where needed, when needed, as needed. The same robotic arm that made a Zaha Hadid flatware set last week can make face shields for medical staff today—ten miles from a hospital in need. No airport required for delivery.

During the second world war the brutality of the war effort had the side effect of revealing the power of modern technologies. Many who had long resisted modernism in architecture and design got used to it during the war, out of necessity; then adopted and embraced modernism out of choice, and without cultural reservations, as soon as the war was over.[37] Likewise, the coronavirus crisis may now show that many cultural and ideological reservations against the rise of post-industrial digital technologies were never based on fact, nor on the common good, but on prejudice or self-interest. More importantly, the global pandemic has also shown that the digital way of making is no longer an option for our future—it is the only viable future we have.

30

Digital design and fabrication have been around—discussed, theorized, tested, and increasingly put to use—for more than thirty years now. Some core features and principles of digital making, as known to date, have been summarized in this chapter. The next chapters will tackle a few more recent trends in design and research. These fall into two main areas: computer-driven robotic assembly and artificial intelligence. Nominally, neither sounds new. Artificial intelligence has been called that way since 1956. Robotic automation was a hot topic in the early 1960s, and industrial robotics was seen as a game changer in the 1970s (and it has been largely adopted since, often without much fanfare, by many mainstream industries, first and foremost the automotive industry). Why are the design professions now suddenly excited by robotics and artificial intelligence, after such a long—almost half a century long—period of neglect? Are we late-comers or trendsetters this time around? This is what we shall try to ascertain in the next two chapters. (A tip-off: robotic automation and artificial intelligence today do not have much in common with what they were half a century ago; even more significantly, today they both kind of work while fifty years ago they mostly didn't.) The last chapter will discuss something that designers do very much care about—and if they don't, they should: what do robotic automation and artificial intelligence *look like*, if we use either, or both, to design and make physical stuff? If something was made by robots or designed by a computer—*does that show*? How? Is there a style of robotic assembly; is there one of artificial intelligence? Finally, some aspects of the political economy of digital making, already anticipated in this chapter, will need some additional considerations, and those will come in the conclusion. Because there will be a conclusion, or a moral, as one is generally offered at the end of most stories. That's why we tell stories in the first place.

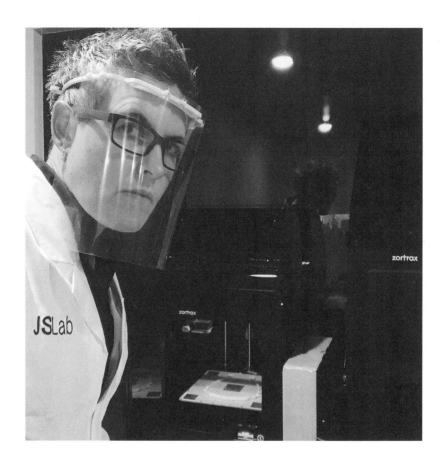

1.1   Jenny Sabin, The Sabin Lab and Cornell College of Architecture, Art and Planning (AAP), Operation PPE, 2020. The Sabin Lab and the Material Practice Facilities at Cornell Architecture, Art, and Planning (AAP) launched Operation PPE with collaborators in Engineering at Cornell University on March 24, 2020, to respond to an urgent request from Weill Cornell Medicine in New York City for PPE face shields. The initiative inspired clusters of makers and designers across the country and internationally. Cornell faculty, students, staff, alumni, industry partners, and home 3D printing enthusiasts delivered over 35,000 face shields to regional and New York City-based hospitals. Courtesy Jenny Sabin and the College of Architecture, Art, and Planning, Cornell University.

2   The Future of Automation: Designers Redesign Robotics

As the name suggests, computational design is based on the use of electronic computers, and it is therefore generally assumed that the rise of CAD has been driven primarily by the technical logics of the computers we use to notate and calculate architectural forms.[1] This storyline can make us forget the role, and influence, of the tools for computer-driven manufacturing and fabrication that some industries first adopted in the late 1950s, and designers started to use extensively as of the mid-1990s. When architects began some serious tinkering with computer-aided design they could not help but realize that the computers we use to draw objects on the screen can not only print out those drawings on paper—hooked up to different peripherals, they can also print out the same objects in 3D, and fabricate them right away. As a result, the integration of computer-aided drawing and manufacturing (CAD-CAM) has been at the core of digital design theory from the very start, and the machinery for numerically controlled fabrication adopted over time has often been as influential and even inspirational for designers as the software at the core of the computer systems themselves.

CNC (computer numerical control) milling machines, a legacy subtractive fabrication technology where a drill moving along a continuous path carves seamless grooves into a solid volume, were the designers' tool of choice during the first age of digitally intelligent design. In the course of the 1990s architects and designers realized that CNC machines—already largely adopted by many industries—were a perfect match for the new, user-friendly programs for digital streamlining (or spline modeling) that were then coming to market. This felicitous pairing, almost a feedback loop, between software and manufacturing tools accounts for the rise in popularity of the smooth and curvy, continuous lines and surfaces that have marked end-of-millennium global architecture—a style that is now often called parametricism. Today the offices of Zaha Hadid and others still use similar software to design splines at colossal, almost territorial scales, but the older among us will certainly remember the time, in the mid- or late 1990s, when every school of architecture around the world wanted a CNC milling machine, which students used to fabricate curvy chairs, spliny tables, and streamlined decorative panels—which, as could be expected, looked more or less the same in all schools and countries around the world.

However, a few years into the new millennium the rise of 3D printing technologies—both cheap personal desktops and industrial-grade machinery—accompanied a significant change of tack in computational theory and design practice: 3D printers often work by materializing small units of printed matter, called voxels; hence the 3D printing of even small and simple objects involves the notation, calculation, and fabrication of a huge number of minuscule, identical boxlike units. At the same time, various "discrete" tools for electronic calculation or simulation, such as Finite Element Analysis, Cellular Automata, and Agent Based Systems, became increasingly popular in the design community—and in design schools. Unlike traditional mathematics, which has developed sophisticated strategies over time to simplify natural phenomena and reduce the number of parameters and variants at play, these new computational tools are designed to deal with extraordinary amounts of unsorted and often meaningless data (now often called Big Data), which electronic computers today can process way better than humans ever could. Not surprisingly, signs of this new and increasingly pervasive technical logic soon started to show up in the work of technically driven experimental designers: raw voxels, for example, were often left visible, sometimes ostentatiously, in numbers far exceeding the powers of human calculation, and at resolutions deliberately challenging the thresholds of human perception. The first known instance of this new computational style—the style of Big Data and discreteness—may have been Philippe Morel's Bolivar Chair of 2004; some ten years later, noted works by Michael Hansmeyer and Benjamin Dillenburger glorified, so to speak, the workings of the industrial-grade voxeljet 3D printers.[2]

These early tools for computational fabrication shared two core features. First, they did not use any mechanical casts, molds, or matrixes to reproduce identical copies. As seen at length in the first chapter, mechanical matrixes have an upfront cost that must be amortized by repeated use; the savings obtained through mass production is called economies of scale. This does not apply to digital fabrication: due to the absence of mechanical matrixes every digital replica is, in a sense, an original, hence making more digital copies of the same item does not make any of them cheaper. Second, however, and mitigating to some extent the import of the above, digital fabrication, well suited for scalability in numbers, appeared for a long time to be eminently non-scalable in size.

Take for example Greg Lynn's seminal *Alessi Teapots*, a non-standard series of ninety-nine variations of the same parametric model, and a picture-perfect demonstration of the technical logic of digital mass customization around the year 2000.[3] Each of the ninety-nine teapots in the series looked like, and was meant to be seen as, a monolith—a single metal block: even if that was in fact not the case, that was nonetheless the idea the series was meant to convey, and this is how each teapot would have been fabricated, if titanium sheets could have been milled, 3D printed, or extruded. However, when trying to apply the same logic to a non-standard series of actual buildings (the *Embryologic Houses*, circa 1997–2001), Lynn was the first to point out that those fabrication processes could not easily scale up.[4] The shell of a parametric, non-standard house could be made of digitally mass-customized, non-standard panels, but these in turn would have to be fastened to a structural frame, and to one another; the frame itself, due to its irregular form, would have been laboriously hand-crafted, and so would each nut and bolt and joint between the parts.

The transfer of non-standard technologies from the small scale of product fabrication to the large scale of building and construction remains to this day a major design issue. Milling machines can mill any number of panels in a sequence, all the same or all different, but the panels must be smaller than the machines that mill them. And most 3D printing happens inside a printing chamber; to print a house in one go, a 3D printer should be bigger than the house it prints.

Some digital makers have tried to duck the issue by building humongous 3D printers; but this technology, often seen as the most suitable for building on the Moon, is still unpractical for building on Earth.[5] As a result, it was long assumed that digital fabrication would be destined primarily to the production of small non-standard components, to be later put together more or less by hand—a process so labor-intensive, and costly, that digital fabrication technologies have been often dismissively seen as suited for making trifles and trinkets at best: teapots, chairs, and experimental architectural pavilions (teapots and chairs requiring little or no assembly; pavilions being mostly assembled by hand by architecture

38

students).[6] This is when architects realized that the ideal tool for the automatic, computer-driven assembly of any number of parts had been in existence for half a century: hiding in plain sight, in a sense—or rather on the factory floor, where industrial robotic arms have been used extensively since the 1960s.

The earliest industrial machines for automated assembly did not have much in common with the artificial humanoids first described by Karel Čapek, the Czech playwright who invented the term "robot" in 1920: industrial robotic arms were, at the start, plain, unromantic, and utilitarian mechanical devices developed to replace wage laborers executing repetitive tasks on moving assembly lines. At the height of the Efficiency Movement the rise of the industrial assembly line had been inspired by Frederick Taylor's theory of task management (1911) and by Frank Gilbreth's time and motion studies (1909). But Progressive Era Taylorism was itself the culmination of a centuries-long effort to reinvent modern work in the absence of skilled labor. Like many cultural and techno-social foundations of modernity, this too had its origins in late medieval Europe.

2.1   Florence, 1450:
The Invention of Notational Work

Life in late medieval and early modern European cities was largely based on, and run by, guilds. Urban guilds, or corporations, regulated and managed all aspects of production, commerce, and often politics; the guilds were also in charge of what today we would call technical education: artisans became masters of their craft and were granted the permission to work independently only after a strictly regulated curriculum of supervised training and exams. At the start of all major building projects (and then again and again during construction) the dominant guilds in the building trades competed for commissions and executed the tasks they received according to best practice, but corporate workers were given vast leeway in the choice of their course of action. There was no design in the modern sense of the term in the Middle Ages, because there were no designers. Medieval master builders were artisans: they conceived the stuff they made as they made it. Buildings made that way were a collaborative, often open-ended endeavor: nobody knew at the start what the final building would look like, nor indeed when building would end;   39 the construction of a cathedral, for example, was often meant to continue forever. That system worked well for a long time, and it produced many masterpieces of medieval architecture. Then some started to challenge it.[7]

The cathedral of Florence was consecrated by Pope Eugene IV on Easter day 1436, and its dome shortly thereafter, on August 30 (the lantern on top was still missing). Almost at the same time young Leon Battista Alberti famously anointed Filippo Brunelleschi as the "architect" of the dome.[8] This was likely the first time since the fall of the Roman Empire—or possibly ever—that the conception and delivery of an entire building was attributed to just one person, and the first time that the term "architect" was used in any modern language.[9] Brunelleschi's struggle to be recognized as the lone inventor of the dome is well known. In the absence of any legal framework that would allow him to establish any right of ownership over his ideas, Brunelleschi could manage to be seen as something akin to the designer of a building only by eschewing all design notations, and making himself indispensable and visible in the flesh, on site, day in day out for all the eighteen years it took to build the dome: distilling instructions only in person and when needed, in the absence of any document or instruction that others could have followed in his absence.[10]

Only a few years later Leon Battista Alberti came up with a better means to the same end—and he wrote a ponderous treatise in Latin to set forth his plan. In Alberti's new humanistic vision architecture is first and foremost an idea, conceived in the mind of just one person—the architect. The entire conception of any new building (what we would call the *design* of a building) must be put on paper before construction starts, precisely inscribed in scaled drawings in plans, elevations, and side views (or something as similar to that as pre-Mongeian geometry then allowed). Construction must then follow the designer's notations in full, faithfully and without any change: the building is the identical copy of the designer's drawings, just like its drawings are a copy of the designer's idea. In the Albertian way of building, design and construction are conceptually and practically separate: all the ideation is on the designer's side, and building is a purely mechanical, servile operation, devoid of any intellectual added value. The Albertian designer doesn't build; the Albertian builders are not allowed to design.[11]

Alberti's treatise, *De re aedificatoria*, was not immediately successful, but Alberti's new way of building by design and by notations must have been in the spirit of the time, as it soon caught on, and Alberti's theory of architecture as an art of design—an authorial, allographic, notational art—invented the architectural profession, and more in general the design professions, as they exist to this day. Medieval building was a craft; Alberti's architecture is a fine art. The medieval builder was an artisan who toiled on site, come rain or shine; the Albertian architect is an intellectual laborer who conceives and composes clean drawings from the quiet of his office and offers professional advice (for a fee).[12] A few centuries later, designers who still earn their living that way have reasons to be grateful to Alberti, for without him they would still be carving wood, laying bricks, and cutting stones—making buildings, instead of making drawings. But what about those on the receiving end of the Albertian revolution? If architecture becomes a notational art (art which is scripted by some but executed by others), building must become notational work: work which is executed by some, but scripted by others.

Alberti's treatise was lengthy by scribal standards, but of its approximately one hundred thousand words only four sentences, all included, refer to human labor, and that is to argue (1) that construction workers should be capable of diligently executing the tasks they are assigned according to the precise instructions they received, finishing on time; (2) that workers should make stuff (*facere*), while architects should think (*praecogitare*); (3) that workers are prone to errors, which can be mitigated by proper use of suitable tools on site; and (4) that workers on site should be constantly monitored by alert supervisors (and here Alberti had to make up a Latin neologism, *adstitores*, from *adsto*, "being there," because he was inventing a new job for which there was no name in Latin).[13] The economic consequences of Alberti's new subdivision of labor are clear: construction workers, who in the medieval guild system were skilled independent artisans working together as associates in something similar to a modern cooperative, become hired hands: centuries before the industrial revolution, proletarians by deeds, if not yet by name. Wage labor was not forbidden in corporate city-states—at the end of the fourteenth century Florence even witnessed what may have been the first proletarian revolution in European history when salaried workers in the wool trades went on a rampage and briefly   41 took power.[14] Brunelleschi himself may have briefly had recourse to foreign, non-corporate wage laborers—that was one of the ploys he was said to have masterminded to break the monopoly that the guilds wielded on the cathedral's building site. But in Alberti's ideological project all work is by definition hired work, because Alberti's deskilled workers are reduced to a brutish labor force—not in the Marxist but in the mechanical sense of the term: animal energy, as we would say today, needed to execute the plans laid out in advance by the Albertian designer.

Alberti never doubted that workers would be capable and happy to work that way. He would soon be proven otherwise, much to his detriment.[15] And apparently, Alberti never suspected that some building materials may be at times quirky or unwieldy, thus requiring some degree of inventiveness, intelligence, and impromptu problem-solving on site. Design adjustments were Alberti's *bête noire*; his was a Laplacian, clockwork universe of absolute human and material predictability—two centuries before the rise of modern science.[16] With hindsight, today we also know that while Alberti's notational mandate was ostensibly meant for, and limited to, architecture and construction work, his new way of making had the potential to be applied to the way we make almost everything. And that's exactly what happened over time: the Albertian authorial paradigm, where ideation is separated from material realization, became a socio-technical staple of industrial modernity. In the Albertian system, as in the modern world in general, thinkers don't make, and makers don't think; the Albertian worker doesn't really work: he executes a script.

2.2   America, 1909–1913: Notational Work Goes Mainstream

Frederick Winslow Taylor (1856–1915), a Philadelphia Quaker, was admitted to Harvard Law School but never enrolled, and chose instead to start a career as a machinist in a factory; he graduated in mechanical engineering by correspondence when he was twenty-seven years old. A wealthy man by birth and earnings, deriving vast income from his patents on steel-processing machineries, he was a keen sportsman and tennis player. There is no reason why he should have ever heard Leon Battista Alberti's name, and nothing in his career and published works indicates that he ever did. Yet, four centuries apart, Alberti and Taylor faced a curiously similar problem. Alberti had to deskill a vast population of highly proficient, trained artisans to make space for his newfangled "authorial" designer; Taylor had to cope with a vast population of unskilled manual labor and invent a new way of working that could use their physical force without training them. He did not have to "deskill" them because they had no skill to begin with, and Taylor's system was designed to keep them that way. The causes and contexts were different, but—at the start of intellectual modernity and at the start of industrial modernity, respectively—the plans of Alberti and Taylor were strikingly alike. Taylor's scientific management brought Alberti's humanistic project to industrial shop work and ultimately to the factory floor.

Taylor did not deny that a body of traditional knowledge existed in each trade, accrued and perfected over time through natural evolution and "survival of the fittest methods." But even long-standing rule-of-thumb practices, he argued, must be tested scientifically and can in most cases be significantly improved by modern time and motion studies.[17] At any rate, regardless of their origin—old or new, traditional or scientific—work practices must be converted into task-based rules, and "enforced standardization of methods" should replace the judgment of the individual workman.[18] This is because, in Taylor's Darwinian worldview, the manual workman is by definition "stupid," for if he were not, he would not be a manual workman.[19] Worse, due to "lack of education" and "insufficient mental capacity,"[20] the manual workman is in fact too inept to independently manage even the most elementary manual tasks; manual workers are ideally "gorillas,"[21] and "their mental make-up resembles that of the ox."[22] In Taylor's experience as an industrial consultant, manual workers proved to be not only irredeemably stupid but often malicious; intent on "soldiering" (deliberately underworking) when not sabotaging machines and resorting to other stratagems to slow down their work:[23] Taylor's human dislike for his workers was on a par with the intellectual disregard Alberti had shown for his, long before. In Taylor's theory the remedy to the workers' ineptitude, hence incapacity to work, is not their education, nor training, but modern scientific management,[24] whereby "the work of every workman is fully planned by the management at least one day in advance, and each man receives in most cases complete written instructions describing in detail the task which he is to accomplish, as well as the means to be used in doing the work."[25] Engineers, Taylor explains, must even tell their gorilla-like workers when to rest and take breaks—as workers, left to their own initiative, would tire themselves to exhaustion.[26] Instructions should be given in writing, via charts and detailed instruction cards.[27] In Taylor's system of task management all work is notational work, as in Alberti's new art of design all art is notational art; Taylor's workers, as Alberti's, execute a program scripted by others. In an unsolicited apology at the very end of his seminal *Principles of Scientific Management* (1911), Taylor protests that his method would *not* turn workers into "automata," or "wooden men."[28] He did not back up his claim.

Taylor's *Scientific Management* relied extensively upon Frank Gilbreth's time and motion studies, first published in 1909, and often cited in Taylor's 1911 book.[29] In a career pattern similar to Taylor's, Frank Bunker Gilbreth (1868–1924) had turned down admission to MIT to start work as a bricklayer; he soon became a prosperous building contractor, specializing in brick masonry. He patented an adjustable scaffold with different levels for bricklayers and tenders, and started to study other ways to improve the efficiency of the bricklaying trade.[30] Gilbreth's plan however was significantly different from Taylor's: rather than systematically eliminating artisanal knowledge, Gilbreth aimed at improving traditional work practices by reducing the number of movements needed for each task. "A mechanic should know his trade," he argued, but workers should be taught to deliver their tasks by fewer, not faster motions.[31] In *Bricklaying System* (1909) he charted the motions needed for laying each brick, and he claimed he could reduce the entire process from a sequence of eighteen moves to four and a half. Workers on the scaffolds could be trained to learn the best moves following footprint charts, similar to floor patterns used to learn dance steps.[32] By repeating this streamlined, optimized choreography on site, bricklayers—and manual workers in general—would acquire over time an "automaticity" of motions that would make their gestures more efficient and precise;[33] thus, Gilbreth claimed, motion studies could deliver better productivity at work than the "separation of planning from performance" (i.e., of design from making) advocated by Taylor's methods.[34]

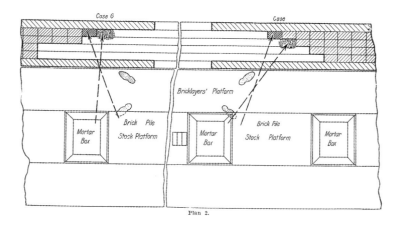

Figure 2.1   Frank B. Gilbreth, *Bricklaying System* (New York and Chicago: Myron C. Clark; London:
Spon, 1909), 153. Motion studies to improve the performance of bricklayers
working on platforms.

Gilbreth was a colorful figure—he had twelve kids, apparently to maximize economies of scale in the cost of their upbringing and education, and he ran his household as a factory. On one occasion he had all family members (himself included) undergo tonsillectomies, one after the other, to show how surgeons could increase their productivity by repeating the same operation many times over in quick succession; Gilbreth himself, who as it happened did not need the operation, was sick for two weeks thereafter.[35] His wife, Lillian Moller Gilbreth (1878–1972), had a PhD in psychology and continued her husband's industrial consultancy long after his death, specializing in domestic economy. Gilbreth had also envisaged to apply his methods to a reform of handwriting,[36] and he explained that his first studies of time and motion had been prompted by the troubles of his own bricklaying business, which was being outcompeted by cheaper "gravel concrete" (cast-in-place concrete walls) and even by construction in reinforced concrete. It appears that at the time of Gilbreth's writing, around 1909, the competition was so intense that bricklayers unions had started to boycott the use of concrete altogether, and unionized workers refused to lay bricks in buildings with concrete foundations or structures of reinforced concrete.[37]

At the start of the twentieth century the advantages of poured concrete walls compared to brick walls were many and patent: ease and speed of delivery, which did not require trained labor; calculability of the structure, due to its continuity and homogeneity; and uniformity of results. Gilbreth insisted that bricklayers should aim at obtaining brick walls as smooth and uniform as concrete walls—but that was evidently a technical non-starter.[38] Any keen observer of technological change around that time would have easily concluded that brickwork was the past, and concrete the future. Around 1912 Thomas Edison built some of his experimental poured concrete houses right in Montclair, New Jersey, a few blocks from where the Gilbreths lived.[39] And when Frederick Taylor, late in life, published the book on scientific management for which he is still famous, he was already known among his contemporaries for a publishing venture now entirely forgotten: starting in 1905 he had published—and then many times republished—a very successful technical treatise on concrete and reinforced concrete.[40]

2.3   Taylor's Reinforced Concrete as a Social Project

Taylor's treatise starts by claiming, without proof, that concrete—both plain and reinforced—is poised to replace brickwork in most building structures.[41] But while the first part of the book presents tried and tested methods and formulas for the making of concrete (or "concreting": proportions of cement, sand, and broken stone; granulometries of the gravel; ratios of water), as well as formulas and tables for determining the compressive strength of plain concrete, the chapter on reinforced concrete opens with a disclaimer: as the theory and design of reinforced concrete are still in an inchoate, elementary stage, the book will present a number of competing theories and different methods, formulas, and tables currently in use in Europe.[42] Taylor's treatise favors the Prussian method, and explains in some detail Christophe's quadratic equation for determining the location of the neutral axis. The tables, however, are complicated by huge variations in the percentages of reinforcement and in the experimental values of the modules of elasticity (E) of both steel and concrete (N, the ratio between the two values, called "r" in the book, varies between 7.5 and 20 but is equal to 10 in some simplified tables). Absent all reference to the mathematical theory of elastic beams, only a few rules of thumb are offered to calculate continuous beams, inserted beams, negative moments, flat plates, and even shearing stress (stirrups and diagonal rods).

This changed drastically in the 1909 edition, where the section on reinforced concrete nearly doubled in size, and was almost entirely rewritten to take into account the endorsement of the Prussian method by the US Joint Committee on Concrete and Reinforced Concrete of 1909;[43] consequently, formulas and tables were simplified, with fixed values for the module of elasticity of both steel and concrete, and N now equal to 15 (as it remained in most countries around the world, till recently).[44] Negative moments and shearing stresses were now explained in more detail, but a short reference to the theory of elasticity was limited to the design of hyperstatic arches. Significantly, in the 1909 edition reinforced concrete was seen as competing against steel structures, not against brickwork.[45]

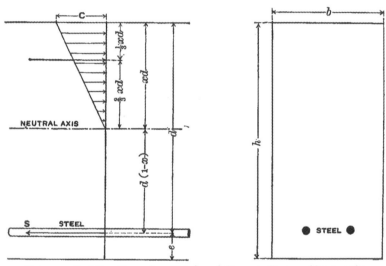

FIG. 90.—Resisting Forces in a Reinforced Concrete Beam. (*See p.* 296.)

2.2   Frederick W. Taylor and Sanford E. Thompson, *A Treatise on Concrete, Plain and Reinforced* (Norwood, MA: Plimpton Press, 1905), 296. Illustration for the calculation of the position of the neutral axis according to the "straight line theory" (Christophe's and Prussian method).

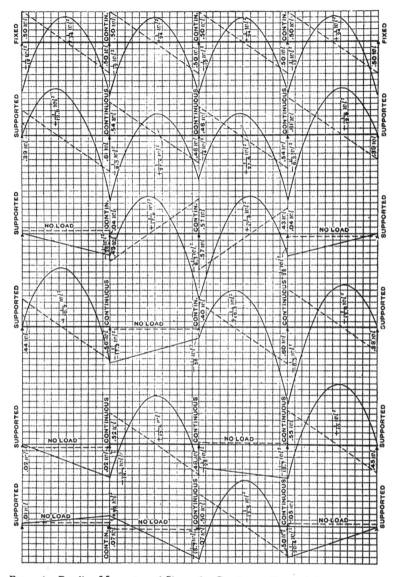

FIG. 136.—Bending Moments and Shears for Continuous Beams, Distributed Loads.

2.3   Frederick W. Taylor and Sanford E. Thompson, *A Treatise on Concrete, Plain and Reinforced* (second edition [expanded] New York: Wiley, 1909), 436. Charts of moments, negative moments, and shearing stresses for continuous beams with different distributed loads on spans and different boundary conditions.

The 1909 edition of Taylor's treatise was often reprinted and updated after Taylor's death (till at least 1939); a comparison between the 1905 and 1909 editions highlights a turning point in the history of the adoption of reinforced concrete, which was still a mostly untested, proprietary technology at the start of the century, when François Hennebique sold reinforced concrete design to third parties based on his company's secret calculation methods. What we still call the "normal" method for the design of reinforced concrete (because it was studied at school until recently) was published by the Belgian engineer Paul Christophe between 1899 and 1902, and by the German engineer Emil Mörsch in 1902; adopted by state regulators in Switzerland (1903), Prussia (1904), and then, in quick succession, by many other countries including the US in 1909.[46] From the point of view of the history of Taylorism, however, and more generally of the history of work and labor in the building and construction industry, it is important to note that Taylor, who copyrighted this book and put his name on the title page, never wrote it.

A footnote appended to the 1905 introduction, and always reprinted in the subsequent editions, attributes "the investigation and study for this book" to Taylor's frequent associate Sanford E. Thompson (1867–1949), an MIT-trained civil engineer. A few chapters are attributed to experts listed on the title page (a specialist on reinforced concrete would be added in the 1909 edition); Taylor himself appears to have been personally in charge of writing a chapter on steel specifications; we must therefore assume that the bulk of the book was Thompson's work, not Taylor's. Yet Taylor—a mechanical engineer, not a civil engineer, who never showed any interest in architectural matters—was evidently the mind behind the whole operation, which he endorsed, authored, and promoted. By publishing the treatise under his name, he brought concrete and reinforced concrete to the attention of the then rising Efficiency Movement. Reinforced concrete, a newly engineered material with almost unlimited promise, seemed well poised to become the "efficient" building material par excellence. Taylor must have also sensed that reinforced concrete in particular was the perfect testing ground for his newly reformed mode of production—and a perfect match for his "scientific" worldview in general.

In 1905 reinforced concrete was a brand-new material. It had no precedent and no artisan tradition. No craftsman, no matter how skilled, would have known what to do with it. The theory of reinforced concrete, in turn, was a product of pure science: the highest mathematics of the time was needed to calculate the tensions taking shape inside the simplest horizontal beams—not to mention more complicated structures. But, unlike steel construction, where most of the parts are factory-made, reinforced concrete is (or was, back then) entirely made by hand, by manual workers toiling on site: the design of the reinforcement—horizontal and diagonal rods, vertical stirrups—must be precisely calculated by the engineers, and all details of the steel rods drawn to scale on technical blueprints, sometimes full-size on paper templates, then forwarded to the foremen. The role of manual workers on site was to execute the blueprints they received—to *copy* the blueprints, in a sense, knowing that every discrepancy, error, or change, could lead to catastrophe. As Taylor did not fail to point out, concrete work is based on a combination of *the least skilled labor* and of *the most alert supervision* in the execution of design notations[47]—thus, as noted, ideally fulfilling a trajectory that started at the very dawn of modernity: as in Taylor's general vision for the ideal subdivision of labor, concrete work does not even require workers—only one engineer on one side, and many automated gorillas (in Taylor's own terms) on the other. In this unprecedented technical environment— inevitable in the case of reinforced concrete work—the agency of the manual worker is reduced to automatic execution, passive stupidity, and brute force. Taylor also noted that the most suitable laborers for such primitive tasks were Italian immigrants[48]—who evidently represented for him the best combination of brainlessness and physical strength; in short, the ideal wage laborer for the new industrial age.

Early in the twentieth century Taylor's project was, simply, *to redesign work—and society—to eliminate skill.* In Taylor's system, *manual workers will do as told, and building materials will behave as expected.* But unintelligent notational work (work that is entirely scripted in advance) can only deal with fully predictable (or, in technical lingo, "rational") mechanical materials, hence Taylor's emphasis on the mathematical theory of reinforced concrete, one of the most successful achievements of classical mechanics: the calculability of reinforced concrete was as essential to his project as the dumb docility of his workforce. Reinforced concrete was the ideal building 51 material for Taylorized notational shop work, just like the moving assembly line would soon become its ideal industrial tool.

2.4   The Automation of Notational Work

In his autobiography, published in 1922, Henry Ford recounts the history of the first moving assembly line, inaugurated on April 1, 1913, for the experimental production of magneto-electric generators in his factory of Highland Park, Michigan. Moving assembly lines for the mass production of the motor, and then of the entire Model T automobile were installed in the same factory between October and December of the same year. Ford recounts that the idea of the moving assembly line came "in a general way, from the overhead trolleys that the Chicago meat packers use in dressing beef"[49]—and that story soon became one of the founding myths of industrial history. Ford does not mention either Taylor or Gilbreth, and he never acknowledged any influence from their theories. For sure, Taylor's and Gilbreth's studies only dealt with the efficiency of traditional manual labor applied to raw shop work, such us gravel shoveling, or brick laying. Yet when setting forth the principles of his new industrial "method" Ford could have been quoting verbatim from either, as the stated purpose of the moving assembly line was not only to "take the work to the men instead of the men to the work," as in Ford's famous tagline; it was to "reduce the *necessity for thought* on the part of the worker" and "reduce his *movements* to a minimum" (emphasis mine).[50]

As per Taylor's plan, an ideally decerebrated assembly-line worker repeats the same scripted motions from sunrise to sunset; as per Gilbreth's theory, the standardization of artisanal gestures is achieved through the repetition of identical motions: tasks based on fewer motions more easily bring about automaticity of motion, hence more efficient motions; to the limit, workflows subdivided into unit tasks each consisting of only one motion will be the most efficient of all. That's what the new industrial assembly line delivered: the economies of scale achieved by the mechanical reproduction of identical parts are notionally matched—and economically compounded—by economies of scale achieved by the machine-like repetition of identical bodily movements. Henry Ford directly referred to Taylor's scientific management when he wrote the entry on "mass production" for the thirteenth edition of the *Encyclopaedia Britannica* (1926)—but only to mark out his difference from Taylor's approach.[51] Taylor wanted his workers to become machines, Ford argued: I use machines instead. Ford was referring to the machinery he used to bring pieces to his assemblers; he did not anticipate that soon a new kind of machine would start assembling those pieces, too—nor did he live to see that.

At the time of Henry Ford's death, in April 1947, his company was in trouble. It had been steadily losing market share since 1930, and many industrial analysts of the time saw it as a goner; General Motors was then the rising automotive company and a celebrated hotbed of engineering and management innovation. Among the new managers that Ford hastily brought in, Delmar S. Harder, hired from GM to become Ford's vice-president of manufacturing, pitched his vision in a speech he delivered in the spring of 1947: "what we need," he claimed, "is more automation."[52] As the term "automation" was a neologism that Harder had just made up, nobody could quite figure out at first what it meant. When it turned out that it meant more machines and fewer workers on the factory floor many started to violently dislike it. On purely philological grounds Norbert Wiener thought the new word was a barbarism and suggested "automatization" instead.[53] Wiener knew a thing or two about making up new words derived from the Greek, and Harder's "automation" could not have had a purer Greek lineage, as it derived from αὐτόματον ("automaton"), neuter of "automatos," meaning "acting of itself." At least that is the occurrence and meaning we find in no less than Homer, who used the term twice in *The Iliad* specifically referring to technical devices: self-opening doors (automatic doors) and self-moving tripods. Homer's technology also included self-driving ships and what we would call human automata, but Homer did not use the term "automaton" for either.[54]

Aristotle refers to Homer's devices in his *Politics*, musing that similar "automatic" machines could eliminate the need for slave labor.[55] In spite of the Aristotelic endorsement, it was mostly from Hellenistic technical literature that the term migrated from Greek into Renaissance Latin and then into modern languages: in 1589 Bernardino Baldi, the Vitruvian scholar and polymath, translated Hero of Alexandria's *Automata* into Italian with the title *Gli automati ovvero macchine semoventi*; a new Latin translation from the original

Greek followed in France in 1701. In the eighteenth century the term "automaton" referred primarily to self-moving statues, either of humans or animals: Jacques de Vaucanson, mostly known for his mechanical duck of 1739, used the term "automate" in his 1738 *Mémoir* to the Royal Academy of Sciences (the French one) describing the mechanism of a self-moving flute player. After the onset of industrialization the semantic field of modern automata reverted to its pristine Homeric meaning: ingenious if inconspicuous technical devices, now often patented (as Westinghouse's "automatic" air brake of 1872). When in 1911 Frederick Taylor claimed that scientific management would not turn a worker into "a mere automaton, a wooden man," he must have been thinking of a puppet-like mechanical body of some sort—a notion for which he evidently could not find the right word, since he used two. The word Taylor would have needed in that context was invented less than ten years later by the Czech writer Karel Čapek.

The story of *Rossum's Universal Robots*, a play that premiered in Prague in 1921 (in New York and Berlin in 1922, with a stage design by Frederick Kiesler; first published in 1920, translated into English in 1923), is well known. Čapek's robots were artificial humans, mass-produced in factories and capable of some intelligence and feelings. They served as slave labor in factories and in farms (the term "robot" derives from a Slavonic word meaning feudal serf labor); as classical slaves, they would find their Spartacus and revolt against their human masters, killing them all (bar one).

As Čapek's robotic parable soon became a popular trope of the techno-friendly Twenties, the terms "robots" and "automata" started to be used more or less interchangeably; in 1942 the science fiction writer Isaac Asimov published his famous "Three Laws of Robotics," less noteworthy for the humanistic principles he aimed to extend to artificial life ("a robot may not injure a human being," etc.) than for giving a name to a forthcoming discipline, "robotics"—a field of engineering dedicated to the design of robots.[56] For all that, the "automation" revolution that Ford and other industrial manufacturers started to promote after the war was not originally seen as related to robotics, either semantically or nominally. Automation was about automatic machines—moving assembly lines, or bottling plants; robotics was about self-moving statuary and artificial humans. The two fields would meet and merge, almost by accident, a few years later.

In 1937 George C. Devol (1912–2011), an inventor and businessman without any engineering training, patented the first and original, Homeric automaton—the self-opening door (albeit he does not appear to have been aware of the precedent). In 1954 he applied to patent a device he called Universal Automation, or Unimation, consisting of a "programmed article transfer apparatus" and a separate device to record the motion-controlling program (originally on an electromechanical relay). The transfer device itself looked like a miniature crane, with a short boom with sliding parts and a pincer, or manipulator, at the end to grasp and move objects. The patent was granted, oddly, only in 1961. Meanwhile Devol had incidentally met (at a cocktail party, it appears) Joseph E. Engelbeger (1925–2015), an electric engineer and a keen reader of Asimov. Engelbeger thought that Devol's machine had legs (metaphorically, as it didn't have any literally); the two founded a company, Unimation, incorporated in 1962, and they started making Unimate machines, which Engelberger decided should be marketed as "robots." It is as a robot, not as a humble "article transfer apparatus" or industrial manipulator that the Unimate captured the imagination of the American public, and the world—in spite of the fact that the machine definitely did not look like a human. In 1966 a Unimate robotic arm featured on the popular *Tonight Show Starring Johnny Carson*, where it poured a glass of beer and performed other stunts (with some legerdemain at play).[57] At the same time Unimate machines were being quietly adopted by a number of car manufacturers around the world, first to carry out dangerous operations (where the robotic arm was more often remote controlled, i.e., not acting automatically),[58] and then to replace manual workers performing programmable, repetitive operations—typically at moving assembly lines. Similar industrial manipulators soon started to be produced by other companies, licensees or competitors, in the US, Europe, and Japan.

Automation, now called industrial robotics, was on its way to revolutionize industrial labor. Since the 1960s the terms automata, robots, automation, and robotics have all been used to describe, with different nuances, human-made devices that look like humans (or animals) or behave like humans (or animals), or carry out tasks on behalf of humans, regardless of what they look like or how they do it. As a result the field of industrial robotics, often seen as a subclass of automation and a highly technical branch of mechanical or

2.4   Unimate industrial robot serving drinks to George Devol, ca. 1962.
From the Collections of The Henry Ford.

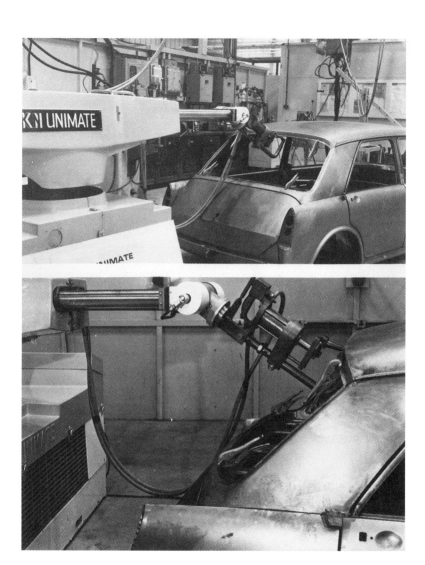

2.5   Unimate industrial robot in use in a car factory, ca. 1967 (apparently a British Motor or British Leyland assembly line). From the Collections of The Henry Ford.

electromechanical engineering, also occasionally borders with more esoteric traditional sciences dealing with the creation of artificial life, and the distant ancestry of today's robotic tools has been stretched to include Paracelsus's Homunculus, Mary Shelley's Frankenstein, the Golem, or Collodi's Pinocchio. This is what Engelberger called the "robotic mystique": robots are automated devices, but with some peculiar features which single them out for special attention.[59]

What these special features are, however, nobody seems to know precisely. The definitions adopted by the International Organization for Standardization (ISO 8373) are confusing and seem at times arbitrary.[60] The original entries, from 1994, had robots down as autonomous, reprogrammable manipulators, where autonomous meant unsupervised; reprogrammable meant capable of performing different operations with the same mechanical body; and the etymology and the meaning of the term "manipulator" both suggest some degree of anthropomorphism, by attributing to robotic machineries at least a hand and likely an arm to append it to.[61] Current ISO 8373 definitions (last updated in 2012) have added some inconsistent automotivity requirements.[62] This may explain why, to this day, a vacuum cleaning device that looks like a round plate on wheels and roams the floor like a puppy is called a robot and a cuckoo clock isn't. The anthropomorphic "mystique" just mentioned may also explain why so much time has been wasted trying to develop humanoid automata meant to carry out tasks well within the reach of plain mechanical automation: a human-looking robot capable of doing our washing is a permanent fantasy of science-fiction writers and technologists alike, but fully automated (programmable) washing machines, in the shape of white metal boxes, have been doing that job since the 1940s—and today they can more or less dry our laundry, too.

Devol's original Unimate patent application from 1954 made it clear that the motions of the robotic arm would be controlled by a system of sensing devices (which Devol called "sensor-based feedback loops"). For example, the mechanism for the extension of a sliding part of the arm would run, when triggered, not for a given number of seconds but until a sensor would detect that the required position was reached.[63] Soon it appeared that sensors and feedback might be equally used to deal with minor incidents in the process, making the machine capable of adapting to some unscripted occurrences; in the "cybernetic" spirit of the time, this adaptivity, or reactivity, was seen as a capacity of the tool to autocorrect, and such autocorrecting devices were then called servomechanisms, or simply servos—in the original technical sense of the term, which is different from the popular one, as often used in the automotive industry, for example.[64] However, the current technical definitions of "adaptive" robotic fabrication did not emerge until the late 1980s, pioneered in particular by John J. Craig, whose handbooks on robotics are still curricular reading for engineering students around the world. In Craig's model, robotic manipulators can tweak the delivery of a scripted task when needed by dint of "non-linear controls of motion," whereby the same coded instruction may result in different robotics movements depending on local feedback from sensors.[65]

As industrial robots were designed to operate in strictly controlled automated environments, robot adaptivity was primarily meant to mitigate such incidents as could result from a limited number of unscriptable factors—an often cited case in point was the trimming and deburring of bindings in arc welds.[66] However, "sensed" robots executing scripted tasks in non-scripted ways—no matter how limited the leeway they were originally granted—are a new kind of technical object; their designers may have been alien to queries on the nature of being, but robots capable of making independent decisions on the fly are de facto new ontological entities. From the early 1960s academic researchers, notably in the then burgeoning field of Artificial Intelligence, did not fail to notice—if engineers did not.

Research in intelligent robotics flourished at MIT and Stanford in particular throughout the 1960s and 1970s. Charles Rosen's famous Shakey, developed from 1966 to 1972 at the Artificial Intelligence Center of the Stanford Research Institute (not to be confused with the Stanford Artificial Intelligence Lab, then headed by John McCarthy) was likely the first self-driving vehicle. A box on wheels that did not perform any task other than plodding its weary way (hence its name) around a cluttered room, it communicated via radio with the bulky mainframe computer needed to process the information it collected from built-in cameras and bump sensors; the robot was in fact remote controlled, but controlled by a computer, not by humans. Also at Stanford, but at the Artificial Intelligence Lab, Victor Scheinman's Arm (1969) was a robotic manipulator designed specifically for automated computer control. Some of Scheinman's technology was eventually acquired by Devol and Engelberger's Unimation (1977) but a later project by Scheinman, entirely based on machine vision, was abandoned due to lack of commercial interest. Similar projects of intelligent, "sensed," and computer-operated robotics (like the Tentacle Arm developed by Marvin Minsky at MIT in 1968) or of autonomous vehicles (Stanford Cart, 1973–1980; Carnegie Mellon's Rover, 1979) did not result in practical applications, nor did they influence the coeval development of industrial robotics.[67]

As late as 1980 Joseph Engelberger, Unimation's co-founder, was not anticipating an imminent market takeoff for intelligent robots. Engelberger's industrial robots were "senseless" machines, trained to repeat the same sequence of regular motions ad infinitum, if needed—or until reprogrammed. The training, or programming, for each robotic task was typically done by humans driving the robotic arm by sight, through a remote control. This sequence of motions was recorded (originally on magnetic tapes) then loaded into the machine so the machine could repeat it identically as many times as needed; unsurprisingly, these kinds of industrial robots were called "playback robots." As Engelbeger admits, "playback" manipulators can only handle pieces "presented to the robotic hand at the same pick-up point and in the same attitude" as when their program was first recorded; in Engelberger's ideal factory this is all the skill a robotic manipulator may ever need, as every random or unscripted factor will have been expunged from the production process before robots are brought in and trained.[68]

Engelberger's 1980 book, *Robotics in Practice*, a comprehensive compendium of industrial robotics, was prefaced by none other than Isaac Asimov. Oddly, Asimov himself—the science fiction writer and noted futurologist—did not envisage any future for intelligent industrial robots. In his view, manufacturing robots are designed to do "simple and repetitive jobs," jobs "beneath the dignity of a human being": robots will shoulder "more and more of the drudgery of the world's work, so that human beings can have more and more time to take care of its creative and joyous aspects." Asimov apparently never imagined that manufacturing—the production of material goods—could itself become a non-repetitive or even creative endeavor.[69]

Almost at the same time as Engelberger's and Asimov's pleas for "senseless" (dumb, repetitive) industrial robotics, Jasia Reichardt, the London-based curator and art critic who in 1968, aged thirty-five, had masterminded the seminal and now legendary *Cybernetic Serendipity* exhibition, denounced and deplored the loss of the imaginative ambition of robotic technologies. Written in the aftermath of the first energy crisis and in the context of the general technological despondency of the mid-1970s, Reichardt's 1978 *Robots: Facts, Fiction and Prediction* tells a story of promises unfulfilled and expectations disappointed: the dream of intelligent robotics, so vivid in the 1960s, had withered in the 1970s, when credits dried up, many projects were abandoned, and research shut down; as Reichardt concludes in the postface to her book, in the 1970s our priorities have changed;[70] innovation in robotics is now more likely to be coming from art than from industry.[71]

Back in the 1970s, at the time of Reichardt's writing, there would have been two main reasons for that. Most projects of intelligent robotics and self-moving vehicles conceived in the 1960s would have required a supply of computing power that at the time was simply not available—and in the 1970s many concluded that it would not be available anytime soon. That was one of the causes of the general fall from grace of Artificial Intelligence—as a branch of computer science—in the mid-1970s. One additional and specific factor, however, played a determinant role in the downfall of intelligent robotics. Industrial automation had been the main market for robotic technologies as of the early 1960s, as well as the driver of most innovations in the field. If early robots were sometimes destined to difficult or specialized tasks—as some still are—massive adoption in manufacturing, particularly in the car industry, only came when robots were brought in to replace blue-collar, assembly-line workers, due to diverging costs of human and robotic labor: a common joke in Europe's industrial capitals after the social unrest of 1968–69 was that robots do not strike for longer paid holidays. Following Taylor's and Gilbreth's principles, assembly-line workers were humans functionally converted into mechanical devices, trained to repeat—ideally—one identical motion, automatically and forever. The dumb robotics machines eventually called in to replace them (i.e., to replace humans already working like dumb robots) would therefore have found themselves in a very congenial environment: a purely mechanical one, devoid of all "non-linear," unscripted reactiveness from the start and by design. Industrial robots simply inherited the intrinsic

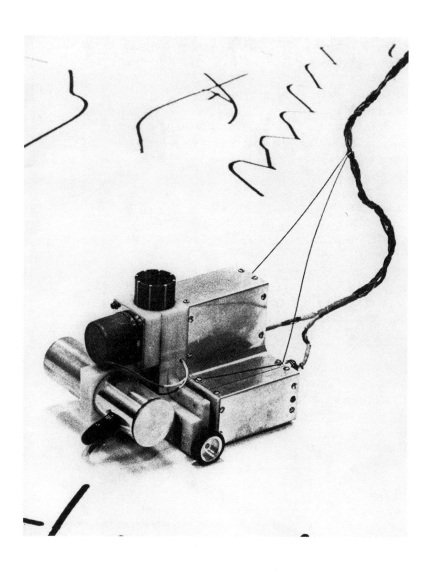

2.6 Jasia Reichardt, *Robots: Facts, Fiction and Prediction* (London: Thames & Hudson, 1978), 60.
Image of Harold Cohen's computer-controlled drawing robot on show at the Stedelijk Museum,
Amsterdam, 1978. Courtesy Jasia Reichardt, London.

stupidity of the human workers they were replacing. Factory owners did not need intelligent robots to replace artificially decerebrated, unintelligent workers—and at the time there would have been no artificial intelligence to power intelligent robots anyway.

2.5   First Steps toward Post-Notational Automation

Two histories of robotics written in the early 2000s tell the story of an ancestral ambition to create artificial humans that unfolded over centuries and accelerated in the twentieth century,[72] flourishing in the 1960s and culminating in the 1970s, when robotic tools were adopted by manufacturers and started to revolutionize industrial mass production.[73] Technical progress in applied robotics didn't stop there, evidently, yet plenty of anecdotal evidence confirms that early in the new millennium the industrial robot was largely seen as a mature, unexciting technology. That was around the time when Fabio Gramazio and Matthias Kohler, young architects then in their mid-thirties, started their seminal experiments with industrial robots in the department of architecture of the Federal Institute of Technology in Zurich (ETHZ). Personal computers of the early 2000s could automatically script most of the instructions, or code, needed for driving the motions of a robotic arm: that was done by some kind of reverse engineering—setting the final positions of the robot's hands, so to speak, then letting the computer calculate all the motions needed to get there. Gramazio and Kohler soon realized that computers could also be tasked to write sets of instructions for regular sequences of different robotic motions, and they concluded that if such sequences could be automatically scripted, bricklaying would be, among the building trades, a prime candidate for robotic automation.[74]

Strange as it seems, it takes a lot of electronic computation to program a robotic arm to lay bricks, because even when building the simplest, straightest, and ideally mortarless brick wall, every brick must still be laid in a new location, next to or above the last brick laid; hence each cycle of bricklaying motions (from pick-up to positioning into place) is geometrically different. Industrial playback robots, programmed to repeat the same motion many times over, could not do that. Today's computer-driven robotic arms can. Moreover, identical robotic motions, and robotic motions that are different from one another within given mechanical limits, cost more or less the same (excluding the cost of some additional lines of code, which is in most cases irrelevant). This means that the simplest brick structure and a more complicated one—a brutally plain, straight, two-mile-long wall and the daintiest, overly ornate baroque vault—come at the same cost per brick laid.[75] Gramazio and Kohler famously demonstrated the architectural potentials of this new technical logic by designing and building a number of geometrically complex brick walls, where the number of calculated variations was intentionally beyond the reach of most skilled artisan bricklayers—assuming that skilled artisan bricklayers still existed and that they had at their disposal an almost unlimited amount of time. More recently, Gramazio and Kohler have applied the same logic of differential, or non-standard, assembly to stocks of different components, each individually designed and custom built: the roof of the Arch Tech Lab at ETHZ, completed in 2016, included almost 50,000 different timber elements automatically assembled in 168 different lattice trusses, covering a surface of 2,300 square meters.[76]

This suggests that, counter to their original technical logic, *industrial* robots can now be tasked to carry out some quintessentially *artisanal* tasks—without any need to standardize the artisanal processes or gestures at play, and without any need to curtail the inherent variability of artisan making. In the case of Gramazio and Kohler's mortarless (dry) walls, all the components were identical or, in the case of their timber roof trusses, all different by design, with each difference notated in full. As the entire process dealt with known entities (parts and places) at all times, each robotic motion could be fully scripted from the start: in ontological terms, Gramazio and Kohler's neo-artisanal robotic variability still resided within the ambit of "linear," notational work. But that could now be about to change.

Given the limited remit of traditional industrial applications, even the rising field of "agile" industrial robotics has till now mostly focused on the robotic systems' ability to respond to handling errors, local incidents or to changes in orders;[77] for example, thanks to machine vision robotic welders can find welding points even when their coordinates are not all known in advance.[78] That may well be all the intelligence needed on an industrial factory floor. But in

2.7   Gramazio Kohler Architects, Gantenbein Vineyard, Fläsch, Switzerland, 2006. Robotic
fabrication of the non-standard brick facade. © Gramazio Kohler Architects, Zurich.

2.8   Gramazio Kohler Research, The Sequential Roof, Zurich, Switzerland, 2010–16.
Digital assembly of a complex roof structure. © Gramazio Kohler Research,
ETH Zurich, ITA/Arch-Tec-Lab AG.

building—as in agriculture, for example, or forestry—dealing with randomness is the rule; predictability the exception. Which is why non-linear, intelligent robotics, mostly superfluous in industrial factories, could be a game changer in the construction industry.

In fact, what may well be missing at this point, and may be slowing down the adoption of adaptive robotics, is not so much the technology as the mentality—our way of looking at design, production, and delivery. For one hundred years we have been trying to convert building sites into factories, so we could mechanize them using industrial tools of mass production. That never worked. Today, we could conceivably do the opposite: we could use post-industrial, adaptive, intelligent robotics to bring efficiency and speed to a neo-artisanal, neo-autographic, post-notational way of making, still capable of dealing with the inherent variability of nature and materials by dint of the inherent intelligence and adaptivity of human making—never mind that the agency now tasked to deliver that human touch may not be entirely human anymore.[79]

Alongside Gramazio and Kohler's experiments with randomized materials or natural materials as found (foams, sand, rubble),[80] systematic research on adaptive fabrication is currently carried out by Achim Menges and Jan Knippers at the University of Stuttgart. Due to Menges's long-standing interest in natural materials (wood in particular) and to his team's more recent experiments with high-tech, non-elastic materials (fiber-reinforced polymers and carbon glass filaments in particular), the pavilions and installations built by the ICD/ITKE Stuttgart team are exemplary case-studies of material-driven fabrication.[81] Explorative testing was traditionally the only way to deal with quirky, lively, or temperamental building materials—materials which engineers used to call "irrational" and now more often called "non-linear." If there is no way to know in advance, by calculation, modeling, or computational simulation, how much weight we can safely load onto the vertical axis of an old and wormy timber pole, for example, experience suggests that we should load it little by little until we start to hear the noise or feel the shivers of some of its fibers cracking. Skilled artisans once knew how to do that—by experience, intuition, or even, some claim, by their inner sympathy with things. Sensing robots today can already emulate some of those ancestral skills—not by magic, but by metrics. Menges has called this mode of adaptive, feedback-driven fabrication "cyber-physical,"[82] and he has duly pointed out that this shift from predefined execution to gradual adaption, or from instruction to behavior-based fabrication, marks the end of the modern separation between design and building, and of the mode of design by notation that was invented during the Renaissance.[83]

Modern design is an authorial, allographic, notational way of making based on calculations and prediction. Its core notational tool, the engineering blueprint, is based on the assumption that all forces and factors at play will behave exactly as expected. To that end—as we have seen throughout this chapter—artisanal work and natural materials alike were laboriously standardized over time, so that industrial workers would repeat identical mechanical motions, and industrial materials could be made compliant with mathematical models (and not the other way around). Steel, a perfect embodiment of the theory of elasticity, is a case in point: the math describing the mechanical behavior of isotropic and homogeneous materials preceded their industrial development in the second half of the nineteenth century. But stones, for example, as found in nature, are all different from one another. This is why modern engineers had to process them to make them bland and tame, homogeneous, predictable, and calculable—thus creating an artificial stone, which we call concrete. Ditto for timber: engineers cannot work with timber as found in nature, because each log is different when chopped off a felled tree; for that reason, the timber we use in building is "engineered" and served in standard and heavily processed formats: plywood, particle boards, laminated timber, etc.

Today, however, intelligent robots tasked to build a wall in an open field could in theory scan the horizon, pick and choose from any random boulder on sight, analyze, combine, and assemble those boulders as found to minimize waste, then pack them together in a dry wall without any need for infill or mortar. Likewise, intelligent robots could theoretically compose with the random shapes and structural irregularities of natural timber—fitting together each log as found, or almost, without having to bring it to a faraway plant, slice it, or reduce it to pulp, then mix it with glue and other chemicals to convert it into factory-made boards with standard measurements and tested structural performance. Before the rise of modern engineering traditional societies lived in a world of physi-

ocratic penury, where manufacturing and building were at the mercy

2.9   University of Stuttgart, ICD, Institute for Computational Design (Professor Achim Menges),
ITKE, Institute of Building Structures and Structural Design (Professor Jan Knippers), ICD/ITKE
Research Pavilion, University of Stuttgart, 2014–15. Sensor driven real-time robot control of cyber-
physical fiber placement system. © ICD/ITKE University of Stuttgart.

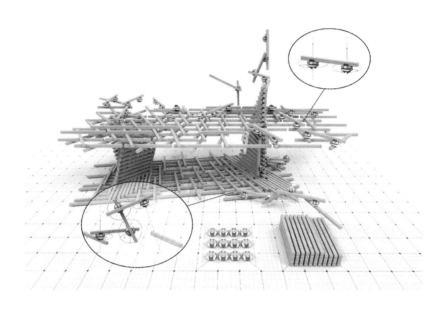

2.10   University of Stuttgart, Institute for Computational Design and Construction, ITech M.Sc. Thesis project 2018, "Distributed Robotic Assembly System for In-Situ Timber Construction." Students: Samuel Leder, Ramon Weber; thesis advisers: Dylan Wood, Oliver Bucklin. Thesis supervisor: Professor Achim Menges; second supervisor: Professor Jan Knippers. © Institute for Computational Design and Construction, University of Stuttgart.

of local supplies of materials and labor. Pre-industrial artisans did not have much choice: most of the time they had to make the most of whatever material, and labor, they had on site. These constraints were eliminated by industrial production, but—for a number of reasons, not limited to climate change—the environmental, social, and thermodynamical costs of mass production and mass transportation are increasingly unwelcome. Today, in compensation, we can use computation and robotic labor to reenact some aspects of our ancestral artisan economy, and recover at least some of its inherent, circular sustainability. Robotic forestry driven by machine vision is already relatively common, and experiments with scan-and-pick stone assembly, and computationally optimized carpentry, made of branches and forks as found, are ongoing.[84]

Another promising line of research in post-industrial, intelligent automation involves the coordination of teams, or swarms, of autonomous robots working together on the same task. Kilobots, a 2014 experiment by the Wyss Institute, is a physical translation of some principles of swarm intelligence that were central to complexity theories of the late twentieth century (on which more will be said in the next chapter). Similar to a cellular automaton, each robotic agent only interacts with its immediate neighbors, and the swarm (in this instance, an aggregate of hundreds of micro-robots) acquires a capacity for collective coordination that complexity scientists call self-organization. Furthering the swarm analogy, the Kilobots designers envisaged teams of self-organizing robots that could be tasked to autonomously build large structures, like colonies of termites and ants building their nests or anthills.[85] A similar project by Gilles Retsin, Manuel Jimenez Garcia, and Vicente Soler (Pizzabot, 2018), specifically meant for building and construction purposes, uses a combination of centralized design and local implementation (adaptivity).[86] Significantly, the Pizzabot robots are identical to the passive construction modules they carry; they can therefore settle anywhere in the structure and be set in motion only when needed. From a distance, the system looks like animated brickwork—a structure made of bricks that move of their own will; true automata in the original, Homeric sense of the term. In another recent experiment by Neil Gershenfeld's team (2019) the modules (in the shape of close-fitting cuboctahedral voxels) are separate from the mobile robotic arms designed to climb on them.[87]

2.11　Kengo Kuma and Associates, SunnyHills Japan, Omotesando, Tokyo, Japan, 2013. Tokyo store for the cake brand SunnyHills. The structure, derived from a traditional technique of Japanese wood construction, is made of slender timber sticks joined in a 3D mesh. Courtesy Kengo Kuma and Associates. Photo © Daichi Ano.

2.12   Kengo Kuma and Associates, Daiwa Ubiquitous Computing Research Building, Bunkyo-ku, Tokyo, Japan, 2012–14. The facade is clad with cedar boards of four different widths, assembled at different distances to provide variations of porosity and transparency. Courtesy Kengo Kuma and Associates. Photo © SS Inc., Tokyo.

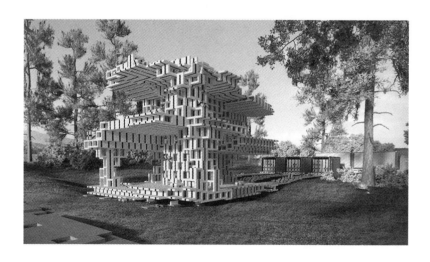

2.13, 2.14　The Bartlett, UCL, Automated Architecture Labs, Pizzabot: Assembler Assembled, 2018. Tutors: Gilles Retsin, Manuel Jimenez Garcia, Vicente Soler. Students: Man Nguyen, Mengyu Huang, Martha Masli, Dafne Katrakalidi, Wenij Wang. A distributed swarm robot in the shape of a pizzabox. The pizzabox-like robots pick up identical passive boxes and climb already assembled boxes to put new ones in place. Courtesy Gilles Retsin.

All the instances just seen point to a general shift of interest of the design community from the *making of parts* to their *robotic assembly*. In different ways, assembly is at the core of recent projects by Skylar Tibbits, Jose Sanchez, Jenny Sabin, Daniel Koehler, Achim Menges, and others;[88] but it is noteworthy that some recent work by the celebrated Japanese architect Kengo Kuma shows evident formal affinities with the robotic assemblage processes being discussed here—without any reference to computational theory, and driven exclusively by Kuma's creative reinterpretation of traditional Japanese craft.[89] Post-industrial, non-standard robotics requires extensive use of computational intelligence, and this will be discussed in the next chapter; chapter 4 will review the visual implications of the ongoing shift from an aesthetics of monoliths to an aesthetics of heteroclites, assemblages, and aggregations. But, to recap and to conclude, the diversity of the design solutions, compositional logics, and modes of agency being envisaged, and here briefly reviewed, should not belie the depth and pervasiveness of the techno-cultural change now under way.

Since the dawn of modernity manufacturing has been increasingly turned into an allographic, notational process, where things are designed by some and made by others. As a consequence, work has been more and more unevenly divided between invention and production, conception and material implementation: Karl Marx decried the elision of the early modern artisan class, replaced by an upper class of capitalist owners of the tools of production (including the ownership of ideas, or intellectual property) and by a lower class of proletarians; Frederick Taylor advocated the elimination of artisan makers, to be replaced by an aristocracy of engineers and a mob of slow-witted, malevolent gorillas. Alberti, the Renaissance humanist, was the first to theorize a design method that would implicitly transform medieval artisans into modern artists—or into wage laborers. Alberti started all this so that he could design buildings to his liking without having to build them with his own hands. Over time, his notational method took over every aspect of modern work. For many centuries that system worked well: notational, allographic work was essential to and inherent in the mechanical technologies powering the modern industrial world. Notational work and mechanical mass production rose together. Today, the two are falling together, because today's post-mechanical computational technologies no longer work that way. Today's robotic work is increasingly intelligent, hence unscripted: neo-autographic, and post-notational. It is also increasingly automated, hence post-human.

Unlike old industrial robots, today's non-standard robots can already imitate some of the inventiveness of intelligent human beings—or at least try to work in similarly inventive, intelligent ways. Tomorrow's intelligent robots will not automate the moving assembly line: they will eliminate it; they will not replace the industrial worker: they will create the automated version of a pre-industrial artisan. One hundred years ago Le Corbusier thought that building sites should become factories; by a curious reversal of roles, today's robotic revolution promises to turn the post-industrial factory into something very similar to a traditional building site. The intelligent, adaptive, "agile" robots that are being developed by the design community are likely the future of manufacturing, but the social and economic import of this technical revolution—unleashed, almost accidentally, by research in computational design and architectural automation—far transcends the ambit of our discipline, and raises questions of greater consequence, and of a more general nature.

3   A Tale of Two Sciences, or The Rise of the Anti-Modern Science of Computation

On a snowy day in the early winter of what must have been 1978, or 1979, I took a walk on a footpath next to the small village, high in the Western Alps, where I used to spend most of my school holidays as a child and teenager. I was then in my second or third year in college and walking with a friend around the same age—I studied architecture and he engineering, hence some occasional disciplinary tension between the two of us. There were a few feet of fresh snow on the ground and we were both walking on snowshoes—the rather primitive snowshoes of the time: ours looked like tennis racquets without the handle, with a hardwood frame holding a web of interlacing strings. It was a laborious walk; and after a while I noticed that my friend's pace looked odd and almost unnatural—as if he hesitated before shifting his weight to either side, applying pressure to each snowshoe in slow motion, in a stunted, almost arrested way. Are you already exhausted or training for ballet dancing, or what, I asked. None of the above, he answered; I am an engineer, so I calculate before I act. In this instance, he continued, my gradual way of applying pressure to each snowshoe little by little, by restrained increments at every step, saves plenty of energy and effort. That is patently wrong, I retorted—after some consideration: at the end of each movement, no matter how you move, your weight will be the same; and assuming the snow is an elastic material, which it may as well be for the limited purpose of our reasoning, its displacement under pressure, hence the sinking of your foot in the snow, will be the same regardless of how you transfer and apply your weight, and regardless of the time it takes for you to do so. I have studied that at school, I think I added: elasticity is a linear phenomenon; if you apply twice the weight, you get double deformation under pressure; apply half the weight, you get half displacement; hence, if you apply the same pressure all at once, or little by little, if you do it fast or slow, or in any order you like, the final deformation or compression of the snow under your weight, hence your effort, will still be the same.

This is why you are not an engineer, he answered; because that's only the first part of the story. Yes, the final compression of the snow may well be the same, but the work I do (hence the energy I spend or transfer to the snow) won't be. That's because work (or energy) is the product of the vertical force I apply (my weight), multiplied by the extent of the vertical displacement of the snow under the weight of my foot; but that displacement is in turn proportional to the force of my weight, hence the deformation work will be proportional to my weight squared. Imagine that I apply my weight (W) to the snow in one go: the work of deformation will be proportional to W^2; then imagine that I apply the same weight in two steps: first, I apply only half my weight (W/2), hence the work of deformation will be proportional to $W^2/4$; assume the second step will be identical to the first: the total work of deformation will then be proportional to $W^2/4 + W^2/4 = W^2/2$; in other words, it will be half the deformation work that would have resulted had I applied the same weight all at once. Now, imagine that, instead of applying that same weight in only two discrete steps, or three or four, I apply it in many sequential steps, each very small; to the limit, imagine the smooth incremental application of a steadily increasing weight or force. In that case, the final result (the final deformation work) will be the summation of an infinite number of infinitesimal amounts. This—he said, drawing the sign of the integral in the snow with his alpenstock—is the magic of differential calculus; you make the sum of many particles that, individually, are worth almost nothing—or, in fact, nothing; but when you add so many, at the end, you get to a real result—an actual number. I cannot write the demonstration in the snow, he continued, so let me skip to the bottom line: assuming you apply every little fractional amount of your weight to the snow only after the preceding amount has been fully absorbed by the compression of the snow under your foot, the final amount of deformation work resulting from the gradual (which we would call "static") application of your weight to the snow will be half the result you would have obtained had you applied the same force (your weight) all at once. Hence by walking the way I walk, smoothly and gradually, with elegance and restraint, I save half the effort; comparing my walk with yours, or with the walk of any ignorant clueless brute who walks in the snow the way you do—without any feeling for snow or understanding of rational mechanics—I can also calculate the amount of chocolate tablets you will need to eat to compensate for the energy, or calories, you waste by being the moron you are. Then he added: and by the way, I didn't invent all this. Someone called Clapeyron did, a polytechnician, while he was a professor at the École nationale des ponts et chaussées, around the mid-nineteenth century; but I could demonstrate the same using the much more elegant principle 81 of virtual works. That, he grinned, was invented in my school—by someone who was still a student when he wrote it.[1]

3.1   The Two Sciences

That is the sermon I got more than forty years ago, at the very outset of my grown-up life.[2] It epitomized, I now realize, many tenets and assumptions, as well as some of the bigotry and hubris of modern science. First, the assumption that snow could be likened to an elastic material would be seen as plain wrong today, yet it might have been somehow warranted back then, in so far as elasticity is a purely mathematical notion, which was being honed around the time Benoît Paul Émile Clapeyron came up with his theorem on deformation work, mentioned above; as there is no elastic material as such, all materials can be seen as more or less elastic, as suited to the partial solution of each matter at hand.

The theory upon which all that rested, the theory of elasticity, was a triumph of modern mathematics—a masterpiece of elegance and ingenuity. It was also the culmination and the perfect image of a modern view of the world. Navier and Saint-Venant, pure scientists in the polytechnic tradition, modeled the deformations of the ideal modern material: a perfectly smooth, homogeneous, isotropic, and continuous slice of inert, mechanical matter. No such material existed at the time when mathematicians first described it, but its notional existence was a postulate of modern science: just like Leibniz's nature never jumps (*non facit saltum*), Saint- Venant's material never cracks. Saint-Venant's material is the perfect ground to run Leibniz's calculus at full speed: calculus describes phenomena that change smoothly over time by incremental variations; these variations, written as continuous functions and graphed as curves, are in turn described by the variations of their tangents (derivatives), and the areas these curves circumscribe can be calculated in a similar way (by reversing the operation). The entire system is based on the idea that every line can be subdivided into smaller and smaller segments, *ad libitum atque ad infinitum*, without ever breaking down. There are no gaps, and no singular points in calculus-based lines; even a simple angle in the graph will jeopardize the whole machine because that's a "jump" (mathematicians would say a discontinuity) that calculus cannot describe.

Baroque calculus invented a mathematical world of smooth, continuous variations; nineteenth-century polytechnicians translated this view of the physical world into the mathematical model of an ideal modern material that functioned in a similarly smooth, continuous way throughout—because that was the material they could best calculate with the mathematics they knew. Since then, modern engineers have been striving to produce a physical material that actually works that way. No natural material does, but steel is a good match: that's a material Leibniz and Newton, Hooke, Young, and Navier would have liked—had they still been alive by the time it was invented: as it happens the invention of the Bessemer converter, which started the mass production of industrial grade steel, was coeval to Saint-Venant's famous *mémoire* on the torsion of solids, which postulated the existence of such a material.[3] One can think of many other artificial materials that, either literally or metaphorically, by chance or by design, fulfilled the modern quest for tame, predictable, smooth, and quantifiable physical continuity. In a sense, this is the material landscape of modernity—in science as well as in technology—and, last but not least, in design and in many visual arts.

The point is, none of this could I have fathomed—or even guessed—in the late 1970s. For at the time, there was no *modern* science for me: there was only one science; that was the only science I knew. Yes, I was vaguely aware that the principles of the science I studied at school had been put into doubt, early in the century, by quantum mechanics and relativity; but those exceptions appeared to be confined to the very small (atoms) or to the very big (galaxies); those being domains I did not plan to inhabit, I was persuaded that most of my daily life, and most of my daily work, would unfold in a Galilean universe of predictable causality and rational mathematical modeling.

Well, that was then. I was made aware of how much our view of the physical world has changed in the space of just one generation when, a few weeks ago, I happened to find myself in the very same location where that first memorable conversation on snow mechanics took place—now long ago. The season being the same, the argument and source of my illumination was, once again, snow; and the reason I was back there—in the snow—is in turn part to my argument. I was there because an unprecedented pandemic had put an end to the world as we knew it, and when the pandemic came many left crowded cities to take shelter in the relative safety of isolated villages—or simply because, with cities in lockdown, all work in remote mode, and all urban life suspended, there was no reason for anyone to be in any city anymore. When, after a few months of urban lockdown, it became clear that the conditions in town were not going to improve anytime soon, I realized that I could, like many others, go live and work elsewhere. I ascertained that the mountain house where I used to spend my holidays as a child—and where I had not been for thirty years—was in fact still standing, very much in its pristine state, but for the addition of a miraculously steady internet connection, courtesy of a big dish-like antenna on the roof and of a private company that beams encrypted data signals from the top of the bell tower of a medieval church down in the valley, a few miles away.

This is why and how, for a few months in the fall and winter of 2020, my wife and I could carry on with our work, full time and relatively effectively, from the middle of nowhere. For better or worse, the pandemic and the internet have brought about a mode of working—and with it an entire way of life—that had been both technically possible and theoretically conceivable for quite some time, yet no one ever seriously thought of endorsing or adopting. However, *pas de rose sans épine*—the very same technical revolution that made that new geography of life and work possible is also to some extent responsible for the rise of a new kind of science, which is quite different from all the science we knew until recently—just as today's world is different from the world of yesterday.

If you live in the mountains, and snow is all you see for many months a year, a new science will make you look at the snow in a new way. That's what I realized when, walking after a similarly abundant snowfall on the same footpath that had been the theatre of memorable conversation I just related, I could not help noticing a bizarre phenomenon taking shape on the rising mountainside on one side of my walk: an array of ubiquitous, minuscule snowballs sliding or rolling intermittently down the slope—many the size of cotton swabs, and many more even smaller, with occasionally a few more sizeable ones here and there, interspersed with the rest. That cascade of micro-avalanches meant, I soon realized, a thaw; and the reason thereof was clear as soon as I stepped above a small ridge, and I was hit in the face by a blast of warm, balmy wind—the föhn, a freak of the capricious Alpine weather, supposedly driving people crazy when it blows. That is unproven, but the föhn does makes temperatures spike, which in turn may make some snow patches unsteady, and likely to slide. Sure enough, I soon heard the ominous, almost chthonian cracking noises exuded by rocks and, one would say, entire mountainsides when they heat up too fast—a worthy source of inspiration for many local folksy tales, and for the art of Charles Ferdinand Ramuz. Soon we heard the first small rocks falling, too, which is when we decided we should turn and walk back home. An actual avalanche came, as expected, later in the evening—it kept to its preferred route and didn't do any harm. So in a matter of few hours the cycle of snow discharge had run full circle, from the cotton-swab size of the first, almost imperceptible little rolling snowballs, to the falling rocks that gave advance notice of an impending and possibly catastrophic major event—the big avalanche that buries people and houses and blows roofs away.

I am certain meteorologists today have tried and tested models to study all this and alert residents when needed. Utterly unrelated, however, this idea of catastrophes as part of a natural continuum that spans all scales—starting with the puniest and proceeding almost seamlessly to bigger and bigger incidents up to occasional major disasters—is at the core of an entire worldview, quintessentially alien and in many ways opposite to the principles of modern science. It is a conceptual framework that has infiltrated many aspects of today's society, technology, and culture, much as modern science used to do; inspiring and driving many of today's global decision makers, as well as many minute decisions we all make in daily life. It is a notional field that goes by many names—chiefly among them, complexity theory. Importantly, if you are a mountaineer, this theory also argues, somewhat counterintuitively, that in the general scheme of things avalanches are a good thing. As a complexity scientist would say, avalanches are the natural way for mountains to self-organize.

Complexity science was not born in the mountains and it did not originally deal with snow; it did deal with earthquakes, however, as seismologists started to notice (as of the late 1940s) that, while big earthquakes are evidently, and fortunately, less frequent than small earthquakes, an eerie and almost supernatural regularity seems to relate the size and frequency of all earthquakes; and they also found out, equally surprisingly, that this regularity of nature could be scripted in a simple formula universally valid across all scales.[4] When graphed, this function translates into an almost asymptotic curve (in the common, not in the projective sense of the term: a curve with two tails that get closer and closer to the two orthogonal axes of a Cartesian diagram without ever touching them). Complexity scientists prefer to graph this function on a double logarithmic scale, so the curve is artificially redressed, so to speak; it then becomes a straight line, and seen from afar it looks like the diagram of a linear phenomenon. Due to this use of exponential scales, complexity scientists also call these laws "power laws."[5] Yet the meaning of these functions is clearer to the layperson if their diagrams are kept in their natural, quasi-asymptotic layout: in common parlance, that would mean that for every million earthquakes with a magnitude of 1, say, there is only one earthquake with a magnitude of 1,000,000 (I made up these numbers for the sake of clarity); but that also means the transition between scales is uneven, because the function rises in value on one axis only when it gets very close to zero on the other axis, with only a small belly between the two tails of the curve. If nature really works this way (and presumably it does, as these laws derive from vast troves of empirical data), we must conclude that nature does not produce many earthquakes of an intermediate, middling scale; just plenty of the very small ones, and very few of the big ones.

Around the time when seismologists were finding these laws in the bowels of the earth, social scientists started to realize that the same laws apply to many human-made systems: to the growth of cities, for example (for every one million cities of size one, there is one city of size one million), or to the use of words in natural languages (for every one million words that are used only once a day, there is one word that is used one million times). Known as Zipf's law, after the Harvard linguist and Germanist George Kingsley Zipf who popularized it,[6] this assumption of an inevitable "loss of the middle ground" in many natural and social phenomena fatally tending towards their asymptotic extremes has been applied to fields as diverse as the Internet (for every one million websites visited by only one person, there is one website visited by one million people), the wealth of nations (for every one million proletarians with a wealth

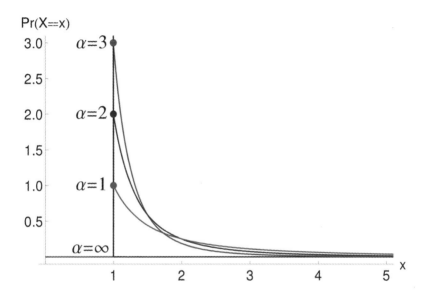

3.1 Power law probability distributions. Example of Pareto distributions for different shape values on a linear (natural) scale. On a double logarithmic scale (log-log) the diagram of a power law becomes a straight line. By Danvildanvil—Own work, CC BY-SA 3.0, https://commons.wikimedia .org/w/index.php?curid=31096324.

of one dollar, there is one capitalist with a wealth of one million dollars), and so on. This latter application of Zipf's law had been prefigured by social scientist Vilfredo Pareto, an engineer by training, whose lectures at the University of Lausanne in 1904 were briefly attended by young Benito Mussolini, then working in Switzerland as a cheesemonger.[7] A mathematical generalization of Zipf's law is also due to Benoît Mandelbrot, whose fractal theory is one of the pillars of complexity science.[8]

The Danish physicist Per Bak must be credited for more explicitly highlighting one of the more momentous consequences of "power law" distributions: if natural systems grow or change by getting farther and farther from their point of equilibrium, and they get more and more unbalanced in the process, sooner or later a major reset will be needed to bring them back to normal. In this view, a system's gradual incremental change is as "normal" as the major catastrophe that the system will need, at some point, to regain its footing. Complexity scientists call this inevitable catastrophe an "emergent" phenomenon, whereby a complex system getting "far from equilibrium" will at some point "self-organize" to recover its pristine state of "organized complexity."[9] Some complexity scientists also believe that during its "critical" phase of re-organization (or self-organization) nature is endowed with some degree of creative agency, vitality, animation, or free will—thus implying that nature can sometimes make independent decisions, like humans or animals. This belief is a mainstay in some areas of complexity theory, but it is improbable, in the etymological sense that it cannot be either proven or disproven—or at least it has not been to date.

Be that as it may, this notion of a quintessentially unbalanced, skewed, blundering world, which finds its way only by never-ending cycles of unsustainable accumulation followed by inevitable catastrophic collapses or explosions, has by now become commonplace—a staple of many current interpretations of natural or social phenomena alike. Per Bak famously thought that complex systems self-organize the way random sandpiles do.[10] As we keep dropping sand on top of a sandpile, the sandpile oscillates between periods of regular growth—which can be modeled and predicted with adequate data—and sudden collapses, when the sandpile creatively reinvents itself, so to speak. The sandpile then finds a new form, and a new structure emerges which, having found a new balance, will then start accruing again, regularly and predictably, till the next collapse. Another suitable image (which Per Bak did not evoke) is that of the coin-pusher machine—of the kind that would have been common in gaming arcades at the time when Per Bak was elaborating his theory.

For those who may not have seen one, a coin-pusher machine consists of a set of sliding horizontal drawers on multiple levels, where coins falling from one drawer to the next randomly pile up. The regular mechanical motion of the drawers pushes the coins over the edge and makes them fall to the lower level; players drop their coins through slots above the upper drawer and hope to collect some when they fall off at the bottom. Most of the time, for every two or three coins dropped into the machine, only one or two get out. The system is then in its homeostatic phase—filling up, in a sense. As players keep playing, more and more coins pile up in small mounds on the drawers, till the miraculous, "critical" moment when one of the piles will become unstable under its own weight and suddenly give way, unleashing a major landslide of coins that will fall into a chute, to be collected by whomever will be playing at the machine at that time. Evidently, the spirit of the game is all about timing the moment of the catastrophe and reaping the maximum advantage from it—to the detriment of former players, as the coins they left in the machine will be the winner's reward (minus the toll extracted by the machine owner). The coin-pusher machine is, I believe, the perfect metaphor for the postmodern theory of complexity. Catastrophes are good because they reset the system and keep it functioning; most of the players at that game will lose a little, but a few will win big.

The epiphanies of complexity theory in the arts and sciences, in politics, society, and economics, are many and diverse. The art historian Christina Cogdell, who devoted a whole book to denouncing some misuses of complexity theory among designers, argues that complexity theory has become something akin to an ideology—what she calls "complexism."[11] That may as well be true, but if so it is worth noting that the influence of complexism extends well beyond the ambit of the design professions. For example, an educator looking at a cohort of students through the lens of complexism may be led to infer that the distribution of student talent follows Zipf's law, hence most students in each cohort or class must be chuckleheads, exception made for the occasional genius. This view of student talent would bring about educational systems very different from those we have institutionalized over the course of the last two centuries. Indeed, most tools currently used for calculating grading statistics tacitly assume that most students are of average talent, with only a few markedly better or worse than all others.

This distribution would be typically graphed on a bell-shaped curve, also known as Gauss curve, or Gauss-Laplace curve, or the curve of normal distribution. In that diagram, which represents the ideal distribution of talent in an equally ideal modern classroom, most students would have close to a median mark, corresponding to the top of the curve; and only a few would find themselves in the tails of the curve, being vastly above or below average. Gauss's curve is, in a sense, a true image of Protestant modernity—the image of a society of which the bulk and core and soul consists of a multitude of equals congregating towards the crowded statistical mean of the social body. That would describe, I presume, the social perception of wealth gaps in egalitarian countries like Denmark, say, or Finland, toward the end of the twentieth century. Zipf's curve, by contrast— the power law curve of unequal distribution—could stand for the distribution of wealth in a country like the United Kingdom in the 1970s: a country conspicuously missing a middle class, where most of the population was equally poor (by the standards of the time), exception made for a few grandees of extraordinary wealth. I cannot support these anecdotal conjectures with any facts and figures (only a specialist could), but Pareto's law was originally meant to describe the allocation of wealth and the distribution of income in Italy and in the UK at the end of the nineteenth century.

The ideology of complexism may also inspire political agendas. A primary corollary of a complexist worldview is the belief that self-organizing systems should be allowed to do what their name portends—namely, self-organize; and that any human attempt to interfere or tamper with natural self-organization is doomed, and ultimately wasteful. Per Bak for example argued that stock exchange crashes are ultimately beneficial because they allow financial markets to reset and in a sense rebuild themselves when necessary, to the general benefit of all. He also argued that random, spontaneous traffic jams are the best way to manage the flow of cars on crowded highways, and he claimed that data and experiments have proven

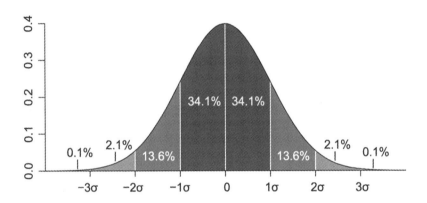

3.2   Example of a normal (or Gaussian, or Gauss, or Laplace-Gauss, or bell) probability distribution curve. By M. W. Toews—Own work, based (in concept) on figure by Jeremy Kemp, 2005, CC BY 2.5, https://commons.wikimedia.org/w/index.php?curid=1903871.

that any attempt to regulate traffic with rules, speed limits, reserved lanes, traffic lights, or the like is ultimately less effective than letting traffic freely find its way (or self-organize), unimpeded, no matter what.[12] Once again, catastrophes are in this instance deliberately chosen as a policy option—the best fix for an apparently intractable systemic problem.

In recent months, this neo-*laissez faire* theory has been tragically tested at a colossal scale. When the coronavirus pandemic struck, countries around the world chose different policies to tackle it. Old-school countries (i.e., countries still imbued with and driven by modern scientific ideas) thought that the virus was, all things considered, a pretty bad thing, and they did their best to eliminate it. Most western countries ridiculed those policies. Their decision-makers and scientists alike—often born and bred complexists of the postmodern or libertarian ilk—knew from the start that the spread of a pandemic is a case study in complexity theory: a pandemic is an emergent, complex phenomenon, and it must be managed with the tools and method of the postmodern science of complexity. Complexity theory teaches that there is no point in trying to inflect the trajectory of a self-organizing system: in the long run, no matter how many countermeasures we take, the inner logic of the system will always prevail. If that is the case, what is our best policy option to cope with an incoming pandemic? Answer: let the pandemic play out. Let the pandemic play out because that's the way complex systems naturally self-organize. Let it play out because the devastation it will inflict will be, in the long run, less costly than any other option.

This was the choice made by most western countries when the pandemic came—with different nuances, as most countries also tried to contain the peaks of the contagion within limits compatible with the capacities of their respective health systems. This is also how most western countries planned their vaccination campaigns: using the vaccine not to eliminate or suppress the contagion (that would be the old modern way) but as a tool to manage its unfolding—the new complexist way. So the two sciences—the old one, based on continuity, causality, the quest for median values, and modern rationality; and the new one, based on unbalances, catastrophes, self-organization and postmodern vitalism—here competed head-on. But in this instance, simple, comparable, and quantifiable results come every day, almost in real time, in the form of body counts and economic costs. At the time of this writing, one of the two sciences is winning—in a big way.

3.2 Modern Architecture and Postmodern Complexity

Architecture was long a supernumerary in the production of the global ideology of complexity, playing only an occasional cameo role. Jane Jacobs, whose seminal *The Death and Life of Great American Cities* (1961) is unanimously seen today as a keystone of twentieth-century urban theory, was familiar with the 1958 Rockefeller Foundation Report, where Warren Weaver had summarized his theory of "organized complexity."[13] Jacobs cited Weaver extensively, and terms and notions of Weaver's theory of complexity contributed to her idea of cities, seen as self-organizing phenomena that emerge from the spontaneous interaction of individual "agents" following rules of behavior solely determined by mutual proximity.[14] Jacobs's point was that cities are made of people, not of buildings, and should be made by people, not by planners (today we would call that a bottom-up approach to urbanism); her immediate targets and foes were the big technocratic plans for urban demolition ("slum clearance") and modernist reconstruction that were then about to destroy many city centers in the United States. Since Jacobs's infusion of elements of complexity theory into urban studies, cities have often been seen and studied as independent, organic phenomena capable of creative self-organization; in more recent times this has incited designers and planners to model cities and city life using cellular automata. More generally, Jacobs's early foray into ideas of self-organization has made cities a primary field of application of complexity theory, and the disciplines of urbanism and planning are to this day deeply indebted to the methodology and ideology of complexism.

Only a few years after Jacobs's plea against modernist cities Robert Venturi's now famous pamphlet against modernist architecture, *Complexity and Contradiction in Architecture*, suggested (by its title, if nothing else) that around that time some idea of complexity may have been gaining traction among the design community. Venturi did cite Herbert Simon's definition of complex systems,[15] but his call for multiple historical, vernacular, or popular references in architecture dealt with visual, figural, or stylistic matters; "complexity" in that context must have been meant in the ordinary, not in the scientific sense of the term. Some notions imported from the sciences of complexity proper may have made inroads among the early techno-friendly designers of the 1960s—of whom Venturi was certainly not one—via their interest in Norbert Wiener's cybernetics. Those few exceptions made (on which more will be said below), complexity science remained mostly foreign to design culture and architectural theory until the surge of computer-aided design and the rise of digital design theory in the early 1990s.

3.3   Architects, Computers, and Computer Science

3.3.1   Before the Digital Turn

Many countries have laid claims to the invention of modern (Turing-complete, reprogrammable) electric computers, of which the first were built toward the end of World War II or immediately thereafter.[16] One of the most famous, the 1946 ENIAC, weighed 27 tons and occupied a surface of 1,367 square feet within the School of Electrical Engineering at the University of Pennsylvania. Computers got smaller and cheaper, but not necessarily more powerful, after the introduction of transistors during the 1950s. Mainframe computers priced for middle-size companies and professional offices started to be available as of the late 1950s, but a mass-market breakthrough came only with the IBM System/360, launched with great fanfare on April 7, 1964. Its most advanced models had the equivalent of 1/500 of the RAM memory we find in most cellphones today (2021).[17] A low added-value professional service dealing with complex problems and data-heavy images and drawings, architecture did not directly partake in the first age of electronic computing: designers were, like everyone else at the time, inspired and excited by the development of the new "electronic brains," as they were called back then, but throughout the 1960s and 1970s there was next to nothing that architects and designers could actually have done with computers in the daily practice of their trade, if they could have afforded to buy one—which they couldn't. Pictures, when translated into numbers, become big files. This is still occasionally a problem today; sixty years ago, it was insurmountable. The first to use computers for design purposes were not architects, nor designers, but mechanical engineers; the expression "Computer-Aided Design," or CAD, was first adopted in 1959 by a new research program in the Department of Mechanical Engineering of MIT dedicated to the development of numerically controlled milling machines. The earliest computer-aided design programs were in fact programs for computer-aided manufacturing.[18]

Around 1963 a PhD student in that MIT program, Ivan Sutherland, wrote an interactive CAD software, called Sketchpad, which used a light pen, or stylus, to draw and edit geometrical diagrams directly on a CRT monitor (a TV screen).[19] Sutherland did not invent the light pen, which had been used by radar operators since the mid-fifties; the novelty of the Sketchpad was a program that allowed for the geometrical definition of scalable planar objects that could be cut, pasted, and resized. These "electronic drawings," however, consisted solely of geometrical diagrams; when the program was shown in Cambridge, England, in 1963, it caused an immediate sensation—but the demonstration was made with slides, because no computer in Cambridge could have run Sutherland's software; even MIT's military-grade computers would have taken hours to recalculate and show each single diagram at a time.[20] Computers of the 1960s were, as their name still indicates, number crunchers; indeed, anecdotal evidence suggests that a handful of major architectural firms that owned mainframe computers at the end of the 1960s had bought them for bookkeeping, not for design purposes.[21] One architectural office that made extensive use of computers, the global planning consultancy of Constantinos Doxiadis, established in 1964 a computer center attached to its offices in Athens, Greece, but Doxiadis's offices around the world used the Athens computer center for statistical analysis and for processing demographic data— not for making drawings, because computers of the time could not draw, and few then imagined they ever would.[22] In compensation some of the loftiest ambitions nurtured by the computer scientists of the time did not fail to inspire some designers, too.

In the introduction to the first edition of *Cybernetics: Or Control and Communication in the Animal and the Machine* (1948) Norbert Wiener recounts how, with a team of scientists and the physiologist Arturo Rosenblueth, he had invented the term "cybernetics" to designate a new discipline devoted to the holistic study of feedback in all processes of communication and control, whether machinic or biologic. The term they chose was derived from the ancient Greek κῠβερνήτης (kubernḗtēs, or steersman: hence the etymology of "governor" in English, or *gouverne* in French, both in the navigational and in the political sense of the term), and it was meant to refer to the steering engines of a ship, seen as the earliest and best-developed forms of autocorrecting servomechanisms.[23] Wiener's cybernetic theory emphasized the similarity between the binary operations of electronic computers and the reactivity of the living cells of the nervous systems, or neurons, which were already known to operate on an all-or-nothing, or binary, mode. This suggested a deeper correspondence between mathematical logic and neurophysiology, warranting the parallel study of computation in electronic machines and of "neuronal nets" in living beings. Wiener's team further grounded the theoretical basis of their new science in a vast program of vivisection of decerebrated cats, carried out at the National Institute of Cardiology of Mexico City.[24] Wiener claimed that his ideas on cybernetics and electronic computing had been endorsed, among others, by John von Neumann at Princeton and by Alan Turing at Teddington,[25] but in the late fifties and early sixties the field of cybernetics was seen as primarily devoted to the study of analog, electromechanical, or organic feedback: so much so that when John McCarthy, Marvin Minsky and others convened the now famous first seminar on Artificial Intelligence at Dartmouth College in 1956 they studiously avoided the term "cybernetics," and it appears they chose to call their seminar "The Dartmouth Summer Research Project on Artificial Intelligence" specifically to avoid any association with Wiener's science and with Wiener himself, who was not invited.[26] When a few years later Minsky wrote a key article, seen to this day as one of the theoretical foundations of Artificial Intelligence, he took care never to use the term "cybernetics"—except in a one-line footnote citing the title of Wiener's 1948 book.[27]

Many years later, for reasons never fully elucidated, the science-fiction writer William Gibson adopted the prefix "cyber" to create the expression "cyberspace," popularized by his best-selling novel *Neuromancer* (1984). Without any direct reference to Wiener's science, the term was soon adopted in popular culture to evoke almost anything related to electronics and computers—up to and including Gibson's own style of fiction, known to this day as cyberpunk; in the course of the 1990s the term was metonymically extended to everything occurring on the Internet, and cyberspace became a moniker for any technologically mediated alternative to physical space. Back in the 1960s, however, the first Artificial Intelligence scientists saw Wiener's cybernetics as something quite separate from the mathematics of computation; even if the analogy between computers and neural networks was generally admitted, the cyberneticians' unconventional and sometimes sulfurous interests in neurophysiology were met with reservations by the engineers and mathematicians that constituted the core of the AI community.[28] To the contrary, many artists and designers, mostly unmindful of the technicalities of early Artificial Intelligence, found the holistic approach and carnal implications of Wiener's cybernetics irresistible.

In the summer of 1968 in London the new age of electronic art was celebrated by the now famous exhibition *Cybernetic Serendipity*. The catalogue of the exhibition included a computer science primer and a lexicon of technical terms; the book opened with a picture of Norbert Wiener (who had died four years earlier) but contained no mention of, nor reference to, Artificial Intelligence. Architecture did not feature prominently in the exhibition, and the few instances of computer-driven architectural design that were shown were notably dull.[29] In this context, the "cybernetic" credentials of the renowned futurologist Gordon Pask should be seen as a sign of his lifelong interest in the interactions between humans and machines, machinic responsiveness and feedback; abundant evidence of this cybernetic line of research can be found in some architectural projects by Archigram and Cedric Price that Pask participated in, or otherwise mentored and inspired.[30] Price's visionary work, in particular, based as it was on modularity, assembly, and mechanical motion, was pervaded from the start by cybernetic ideas of computer-controlled, reconfigurable buildings.

As Price didn't leave blueprints for his most famous projects, we do not know precisely how computers would have managed to move and reposition the modular components that were plugged into the vast steel frame of his famous Fun Palace (1963–67); similar ideas reoccur elsewhere in Price's work, but it is in his later Generator Project (1976–79) that we find a fully developed attempt at the computerized governance of an entire built environment (a theme park that should have been built on a plantation in the southern United States). All the installations in the park would have resulted from the recombination of a set of 150 modular cubes, each the size of a small room, to be permanently moved around by cranes following users' feedback or recalculations by a central computer. Price appears to have claimed that his Generator Project was the world's first intelligent building,[31] but we know today of at least one very similar precedent—Nicholas Negroponte's SEEK installation of 1970, where cubes were rearranged inside a big, aquarium-like box by a robotic arm driven by a computer that interpreted, somehow, the intentions of a population of gerbils. Similar modular boxes were also the basis of Negroponte's URBAN2 and URBAN5 interactive design systems, all illustrated in Negroponte's 1970 book, *The Architecture Machine* (sans gerbils, which were added as the free-will ingredient—the human factor in the cybernetic machine, so to speak—only in a show at the Jewish Museum in Boston, titled *Life in a Computerized Environment*).[32]

In spite of and unrelated to Negroponte's transient interest in self-organizing rodent communities, his groundbreaking *Architecture Machine* is generally seen today as the first fully developed attempt to apply the new science of electronic computation to architectural design. His team, the Architecture Machine Group at MIT, imagined a computer that would act as a universal design assistant, capable of empowering end-users (customers) to design houses on their own. The machine would have guided the customer through the entire design process via a dialogue of formalized questions and answers: the program queried the customer and offered design options at each step, based on the customer's replies. All the necessary design knowledge was inscribed in a set of rules and models pre-installed in the machine—which is why these kinds of Artificial Intelligence engines are known to this day as expert systems, or knowledge-based systems. The customer was supposed to start the game without any prior design expertise; no architect would be needed at any time during the design process. The story told in the book is that the machine was built and tested, and it didn't work.

When it appeared that Artificial Intelligence could not deliver machines capable of replacing designers, and that cybernetics might not deliver buildings capable of redesigning themselves, architects started to lose some of their pristine interest in electronic computers.

They were not alone: in the early 1970s the larger research community was having some serious second thoughts on the viability of many projects of early computer science. As disillusionment set in, funding and research grants (particularly from the military) soon dried up; this was the beginning of the period known in the annals of computer science as "the winter of Artificial Intelligence."[33] For a number of reasons, political as well as economic, amplified in turn by the energy crises of 1973 and 1979, the social perception of technology quickly changed too; as the new decade progressed, the almost unbounded technological optimism of the 1960s gave way to technological doom and gloom. While computer science went into hibernation (and many computer scientists found new jobs in consumer electronics), architecture's fling with the first age of electronics was quickly obliterated by the postmodern leviathan. In architecture, PoMo technophobia relegated most cyber-theories and artificial intelligence dreams of the 1960s to the dustbin of design history. Throughout the 1970s and 1980s, while architects looked the other way, computer-aided design and computer-driven manufacturing tools were being quietly, but effectively, adopted by the aircraft and automobile industries. But architects neither cared nor knew that back then—and they would not find out until much later.

Futurologists of the 1950s and 1960s had anticipated a technological revolution driven by bigger and ever more powerful electronic computers. To the contrary, when the computing revolution came, it was due to a proliferation of much smaller, cheaper machines. These new, affordable "personal" computers could not do much—the first, famously, did almost nothing—but their computing power was suddenly and unexpectedly put at everyone's disposal, on almost everyone's desktop. The IBM PC, based on Microsoft's disk operating system (MS-DOS) was launched in 1981, and Steve Jobs's first Macintosh, with its mandatory graphic user interface, in 1984. The first AutoCAD software (affordable CAD that could run on MS-DOS machines) was released by Autodesk in December 1982,[34] and by the end of the decade increasingly affordable workstations (and soon thereafter, even desktop PCs) could already manipulate relatively heavy, pixel-rich and realistic images. This is when architects and designers realized that PCs, lousy thinking machines as they were—in fact, they were not thinking machines at all—could easily be turned into excellent drawing machines. Designers then didn't even try to use PCs to solve design problems or to find design solutions. That had been the ambition of early Artificial Intelligence—and that's the plan that had failed so spectacularly in the 1960s. Instead, in the early 1990s designers started to use computers to make drawings.

They did so without any reference to cybernetics or computer science. They looked at computer-aided drawings in the way architects look at architectural drawings—through the lenses of their expertise, discipline, and design theories. This is when, after the false start and multiple dead-ends of the 1960s, computers started to change architecture for good. By the time that happened, the intellectual project of early computer science had been not so much demoted as entirely forgotten; when a new digitally driven architecture started to emerge, at the end of the 1990s, designers were reading Gilles Deleuze and D'Arcy Thompson, not Norbert Wiener or Marvin Minsky. And none of the protagonists of the first digital turn in architecture ever thought for a minute of the Centre Pompidou in Paris—a belated avatar of Cedric Price's Fun Palace, and the main built testament to the cybernetic dreams of the 1960s—as a precedent or source of inspiration.[35] Actually, quite the opposite.

3.3.2 The First Digital Turn

The story of the first digital turn in architecture is, by now, relatively well known; often recounted by some of its protagonists, it has recently become the object of the first historiographic assessments. At the end of the 1980s many schools of architecture in Europe, the US, and Canada offered some basic training in computer-aided design. CAD was taught as a run-of-the-mill device for making cheaper, faster drawings—a mere drafting tool. But then some started to notice that computers, alongside new ways of making drawings, also offered ways for making new kinds of drawings—drawings that would have been very difficult or perhaps impossible to make by hand. A new theory of "digital design" then emerged, and architectural shapes began to change, as well as the culture and the economics of design and construction at large.

In the early 1990s a group of young designers interested in new technologies met in or around the Paperless Studio, created at the Graduate School of Architecture, Planning and Preservation of

Columbia University by its Dean, Bernard Tschumi.[36] One of them, 29-year-old Greg Lynn, in the spring of 1993 guest-edited an issue of *AD*, "Folding in Architecture," which is now seen as the first manifesto of the new digital avant-garde.[37] The "fold" in the title referred to the title of a 1988 book by Gilles Deleuze, and Deleuze's bizarre infatuation with the smooth transition from convexity to concavity in an S-shaped curve (which he called "the fold," and onto which he bestowed vast mathematic, aesthetic, and art historical significance) would prove contagious, as it came to define the shape, or style, of digitally intelligent architecture—in the late 1990s and, to some extent, to this day.[38]

Deleuze's aura, formidable as it was at the time, could not alone have managed that. A concomitant and possibly determinant cause was the arrival on the market of a new family of CAD software, which allowed the intuitive manipulation of a very special family of curves, called splines. Hand-made splines had been used for centuries for streamlining the hull of boats, and spline-modeling software had been widely adopted by car and aircraft makers since the 1970s for similar aerodynamic purposes. New, affordable, and user-friendly CAD software that became available as of the early 1990s put streamlining within every designer's reach,[39] and digital streamlining soon became so ubiquitous and pervasive that many started to see it (wrongly) as an almost inevitable attribute of digital design. Mathematical splines are continuous, differentiable functions, and through one of those felicitous blunders that often have the power to change the history of ideas, spline-based, aerodynamic curves and surfaces then came to be seen as the digital epiphany of the Deleuzian Fold. Gilles Deleuze himself, who died in November 1995, was not consulted on the matter. For what we know of him, chances are that he had very little interest in car design, streamlining, and aerodynamics in general.

Despite its recent wane, the popularity of digital streamlining—in the 1990s and often beyond—has belied and obfuscated the importance of the parametric technology underpinning it, which was and is still quintessentially inherent in the digital mode of production, regardless of form and style. Another idea that Gilles Deleuze had devised (in collaboration with his gifted student, the architect and polymath Bernard Cache) would prove crucial for digital architecture: scripted code for digital design and fabrication often uses parametric functions—functions where different parameters or coefficients can generate an infinite range or "families" of objects that share nonetheless the syntax of the original function and which, due to this commonality of code, may share some visible features, and look similar to one another. Deleuze and Cache called this new type of generic technical object an Objectile, and Deleuze's and Cache's Objectile remains to this day the most pertinent description of the new technical object of the digital age. Greg Lynn and others soon came to very similar definitions of continuous parametric variations in design and fabrication.[40]

This was the mathematical and technical basis of what we now call digital mass customization, as discussed at length in the first chapter of this book: a new, non-standard mode of digital design and fabrication, where serially produced variations do not entail any supplemental cost. Digital mass customization is not only a fundamentally anti-modern idea—as it upends all the socio-technical principles of industrial modernity; it is also the long-postponed fulfilment of a core postmodern aspiration: a postmodern dream come true, in a sense, courtesy of digital technologies. Digital mass customization may deliver variations for variations' sake—in theory, at no cost. This is a plan that Charles Jencks could have endorsed—and in fact he did, as we shall see hereinafter; and, in so far as it is animated by a powerful anti-modern creed, this is a conceptual framework that Gilles Deleuze would have found congenial—and indeed, excepting the technological part of the plan, which was mostly due to Bernard Cache, this is an idea that can be traced back to him.

By their own technical nature Deleuze's and Cache's Objectile, and digital parametricism in general, are generative tools: each parametric script contains potentially an infinite number of different instantiations (or, in philosophical terms, "events"); each set of parameters in the script generates one variation. Borrowing a line from medieval Scholasticism, we could say that parametric notations are *one in the many*; events generated by them are *one out of many*. Therefore, from a design perspective, the scripting of a parametric notation is always, conceptually, the first step of a multilayered design process awaiting further completion. Sometimes the author of the original script is happy to delegate some of its subsequent fine-tuning to others, thus inviting others to participate, to some extent, in the design process, and indeed as of the early 2000s participatory and collaborative practices have emerged as a mainstay of digital design and of digital culture at large.[41] Yet alongside the pursuit of these fairly traditional modes of collective or even anonymous (crowdsourced) authorship, digital designers at some point realized that the same technical premises would theoretically allow them to devolve some design choices to non-human co-authors—to intelligent machines. Computers in particular can conceivably be tasked with testing many random variations in a sequence, then choosing one or more based on defined criteria, guidelines, or benchmarks. This would not be unlike a normal process of computational optimization, of the kind that is routinely carried out in experimental sciences. In design theory, however, this machinic approach proved a fertile ground for the introduction of many ideas derived from complexity theory, and the resulting conflation of complexism and computationalism has since held sway over many areas of experimental architecture.

3.3.3 First Encounters between Complexism and Digitally Intelligent Design

Complexism's first inroads into digital design theory came via biological and evolutionist metaphors. Designers had long been familiar with the work of Scottish biologist D'Arcy Thompson (1860–1948). This was primarily due to the popularity of his drawings, showing the long-term effect of external mechanical forces on the formation of living organisms (think of flatfish living at the bottom of the sea, like a turbot, or a sole, being flattened out by the water pressure). Thompson's gridded diagrams described anatomical transformations as if flesh and bones followed laws of formal congruence and material elasticity similar to those used in structural engineering, and for that reason Thompson's drawings have always been popular with designers; early in the 1990s the diffusion of the first tools for digital image processing added to the appeal of his pre-digital "morphing" bodies.[42] This eminently visual fascination with Thompson's drawings did not originally elicit any specific interest in his morphogenetic theories. But code is code, and the analogy between parametric codes and biological ones was already common among digiterati by the late 1990s (its wider consecration came only a bit later, through exhibitions like Foreign Office Architects' *Breeding Architecture* in 2003 or the *Gen(H)ome Project* of 2006).[43] Parametric codes, as well as genetic ones, are invisible scripts that produce a variety of visual manifestations. Following from this analogy, scripted mathematical code came to be seen, metaphorically, as "genotype," and the forms generated from it as "phenotype"; likewise, iterative methods of mathematical optimization, of the kind mentioned above, started to be described in genetic and evolutionary terms.

The metaphor is certainly pertinent: heuristic problem-solving proceeds by trial and error (random variations and selection of the best, or "fittest," among the many solutions tested)—in this respect, not unlike the evolution of species according to many Darwinian and post-Darwinian theories. As a consequence, digital designers have adopted an entire lexicon that computer science has in turn imported from theories of biological evolution: a milestone in this story was John Holland's definition of genetic algorithms in his seminal *Adaptation in Natural and Artificial Systems* of 1975;[44] computational optimization is to this day often described as a self-organizing, emerging phenomenon, and mathematical algorithms as evolutionary notations, capable of improving over time by spawning random variations, of which the less fit will be discarded.[45] The most pertinent and legible experiment of computer-aided design based on genetic algorithms, almost a demonstration of methods, remains to this day John Frazer's *An Evolutionary Architecture* of 1995.[46]

Also in 1995 Charles Jencks published a strange and, back then, largely misunderstood pamphlet, *The Architecture of the Jumping Universe: A Polemic* (subtitle: *How complexity science is changing architecture and culture*).[47] The book was a primer of complexity science—a simplified introduction to complexism specifically meant for the design professions, interspersed with forays into cosmology and other spiritual arcana that were then becoming a major inspiration for Jencks's own work as a designer and landscape architect. Jencks's explanation of the core tenets of complexity was brilliant and it still reads well today, in spite of Jencks's sometimes overzealous attempts to translate every scientific theory into recognizable visual forms. Jencks, who ostensibly knew nothing of complexity sciences in 1977, when his book *The Language of Post-Modern Architecture* spearheaded postmodernism in architecture and in the humanities at large, could convincingly argue in 1995 that complexity sciences vindicated his own idea of postmodernity, and that architectural postmodernism and the sciences of complexity had always been part to the same view of the world—one which rejects the modern science of causality and determinism as well as the technical logic of mechanical, industrial modernity. In fact, Jencks saw complexity as the manifestation of a new postmodern science at large, one that included quantum mechanics, indeterminacy, non-linearity, chaos, fractals, emergence, system theory, and more (notably, this list did not include post-structuralist philosophy, in spite of its determinant influence on architectural theory); Jencks's reference to "the post-modern sciences of complexity" as a whole may have been biased and a bit self-serving, but it struck a chord, and it has stuck.

Given this all-encompassing approach, Jencks's examples of "post-modern complexity" in architecture in the 1995 edition of the book were disparate, ranging from Bruce Goff to Rem Koolhaas, and from Richard Rogers to Andrés Duany. This changed with the revised edition of 1997, which appears to have been hastily reprinted with the addition of a post-script that refers to Foreign Office Architects' winning entry for the Yokohama Port Terminal Competition (1995) and to the recent inaugurations of Peter Eisenman's Aronoff Center (Cincinnati, Ohio, 1996) and Frank Gehry's Guggenheim Bilbao (1997).[48] Here Jencks, as if struck by an illumination, seems to have come to the conclusion that digital technologies for design and fabrication are the *deus ex machina* that will allow postmodern design ideas and the new science of complexity to coalesce, merge, and deliver a new non-standard, mass-customized, post-industrial, non-linear architecture (Jencks, who knew full well what "non linear" means in mathematics, seems to have at times played games with the ordinary sense of the term).

Jencks had shown occasional interest in electronic computation in the 1960s, when he was a student of Reyner Banham; much later in his life, his Damascene conversion to digital technologies signaled the convergence of complexism and computationalism in the mid-90s, when a new generation of forward-looking, experimental architects embraced computer-aided design under the influence of a powerful postmodern impetus, and further absorbed anti-modern ideas of complexity from a variety of sources, including computer science. Jencks would remain faithful to his new 105 creed to his last, endorsing and promoting the work of many young digital creators, to the dismay and disbelief of many of his former PoMo acolytes.

On the other side of the Atlantic Sanford Kwinter's eloquent column "Far From Equilibrium," published in the New York magazine *ANY* from 1995 to 2000, introduced topics and themes from complexity science into American architectural discourse, and made them popular among some of the most influential architectural thinkers and opinion-makers of the time. Kwinter endorsed a vitalist and at times libertarian or nihilistic view of complexity, but his interest in what we now call the digital turn in architecture, which was then in the making, appears to have been limited; only one of his *ANY* columns was entirely devoted to computation, and only to deplore the preeminently bland and technocratic adoption of computer-aided design in mainstream architectural practice. Instead, Kwinter called for a fully-fledged complexist revolution—a leap that would hijack electronic computation and unleash its potential to "free us from the multiple tyranny of determinism and from the poverty of a linear, numerical world" by tapping into "the rich indeterminacy and magic of matter" and pressing toward "the archaic world of natural intelligence . . . and the dark, possibly unfathomable mysteries of nature itself."[49] With historical hindsight, we now know that this is exactly what happened. For better or worse, complexism did hijack digitally intelligent design.

3.3.4 The Emergence of Emergence in the Early 2000s

By the end of the 1990s the postmodern science of complexity was ready for prime time, and complexism soon pervaded popular science, as well as popular culture: Steven Johnson's best-selling *Emergence: The Connected Lives of Ants, Brains, Cities, and Software* (2001) firmly established the legitimacy of an ideology that claimed to hold a key to our understanding of natural phenomena, social phenomena, and technical networks alike; and which furthermore saw the concomitant rise of the new digital technologies and of the so-called digital economy as the fulfilment and vindication of a law of nature predicated on the innate capacity of all complex systems to self-organize by dint of cycles of growing unbalances and catastrophic disruptions. By that time, the financial markets and economics in general were commonly seen as a primary field of application of complexity theory: Paul Krugman's first best-selling book, *The Self-Organizing Economy*, which also included a primer to complexity theory in general, was published in 1996. Even if that was likely not Paul Krugman's intention, many arguments from his book were quickly embraced by neo-liberal and libertarian economists. Artists and designers around the turn of the millennium were also familiar with a more esoteric version of complexism infused in the work of theoretician and filmmaker Manuel DeLanda, whose influential *A Thousand Years of Nonlinear History* had been published by Zone Books in 1997.

Christina Cogdell has convincingly argued that the Emergent Technologies and Design (EmTech) master's program at the Architectural Association School of Architecture (AA) in London was the crucible where, in the first decade of the new century, a creative conflation of complexism, computationalism, and architectural theory was successfully transmuted into a conceptual framework that was both relevant and hugely appealing to cutting-edge computational design research; and indeed the idea of emergence that was forged at the AA during those years soon spread to the computational design community around the world, and it remains to this day (2021) one of the most pervasive and influential design theories of our time—one that has already left an indelible trace on design history and on the history of architectural education around the world.[50] The Emergence and Design Group was founded in 2001 by Michael Weinstock, Michael Hensel, and Achim Menges, and the group authored several issues of *AD*, of which the first, "Emergence: Morphogenetic Design Strategies," published in 2004, is reportedly the best-selling issue in *AD* history.[51] Important books were published by the trio, and by Weinstock alone, in 2010;[52] Hensel and Menges left the AA around that time and the program, which still runs, was until recently led by Weinstock. Christina Cogdell, who was a student in the EmTech program in 2011, has devoted pages of scholarly analysis to the ideas of Weinstock in particular. In her view, Weinstock's teleological view of complexity theory leads to an almost irrational belief in the palingenetic power of emergence: if complexity is the most general, universal law of nature, and the laws of complexity mean that all systems, natural and cultural alike, are predestined to build up always more intolerable levels of unbalance till the inevitable catastrophe from which a better (i.e., even more unstable) system will emerge ("evolution by jerks"), then all human action is irrelevant, and all catastrophes are good.[53]

Unrelated to the above, but at the same time, a new strain of postmodern scientific ideas irrupted into design culture from an unsuspected quarter. Stephen Wolfram, a mathematical prodigy and entrepreneur, wrote three books on particle physics at the age of fourteen and in 1981, aged twenty-one, was the youngest recipient of a MacArthur fellowship. By the late 1990s many designers were familiar with his software Mathematica, which they used for calculations but also for image processing and mathematical visualization. Mathematica's original purpose was to offer a user-friendly interface to process the kind of mathematics we learned at school, using familiar notations—but at computer speed (it does much more than that now). When, in 2002, Wolfram published *A New Kind of Science*, a monumental handbook on Cellular Automata, many in the design community were intrigued.[54]

Cellular Automata are, at their simplest, systems consisting of a grid populated by identical cells in two different colors (either black or white), and of a set of rules that determine the transformation of the color of a cell (from black to white or back) based on the color of the neighboring cells. Rules can be applied iteratively for as many times (or steps) as needed, creating a different configuration of the system at each step. There are no shortcuts to help us predict how each rule will play out: the only way to know what the system will look like after any number of steps is to let the game run till the desired step has been reached. But charting each transformation by hand, cell by cell, albeit perfectly feasible, would take an inordinate amount of time: a post-human scientific tool, Cellular Automata are made to measure for the processing speed of electronic computers, and would be of little practical use without them. Conversely, by letting computers run the game at full throttle, Cellular Automata can easily simulate complex natural phenomena that modern science has traditionally seen as indeterminable or not calculable, such as the growth of crystals, the formation of snowflakes, the propagation of cracks, or turbulences in fluid flow. Wolfram could thus argue that Cellular Automata are a new science (literally, "a new kind of science") suited to modeling complexity—but that reasoning may be tautological. Cellular Automata are construed according to the very same rules that define complex systems. Cellular Automata are not, strictly speaking, a model of complexity (even if they can be used as such) because they are themselves a complex system by definition. They are, so to speak, the thing itself; they were designed that way.

The concept of Cellular Automata derives from studies by John von Neumann and Stanislaw Ulam at the Los Alamos National Laboratory, from 1947 to the late 1950s. Wolfram remarks that von Neumann and Ulam hoped to convert the results of their experiments with cellular automata back to differential equations,[55] thus implying that cellular automata back then were not yet seen as a fully-fledged alternative science. Regardless, Warren Weaver's definition of organized complexity (from his articles of 1947–48 and his better known 1958 Rockefeller Foundation Report) reads almost as a verbatim description of Cellular Automata, which is even more surprising as Weaver never mentioned Cellular Automata,[56] and conversely, it does not appear that the mathematicians at Los Alamos ever acknowledged any influence on their work from Weaver's then inchoate theories of complexity. We must therefore assume, pending further studies, that Cellular Automata and the new science of organized complexity emerged at the same time, independently, but giving voice to a very similar view of nature, society, and science. Early in the new millennium, when architects started to play with cellular automata as a practical design tool, they soon found out that cellular automata were the perfect instrument and partner for a truly complexist strategy of computational design. That was a marriage made in heaven; but one that had been in the making for almost half a century.

3.3.5   The Second Digital Turn

Alisa Andrasek, Philippe Morel, and Neri Oxman were among the first computational designers to experiment with Cellular Automata, then mostly known via Wolfram's 2002 compendium.[57] As mentioned in the previous chapter, early in the new millennium newly affordable additive fabrication technologies, such as 3D printing, showed that objects could be created by piling up stashes of independent, homogenous units or blocks, called voxels by analogy with the pixels that compose digital images. That mode

3.3　Biothing, Alisa Andrasek, The Invisibles, exhibited at Prague Biennale of 2003. Cellular Automata used for generating sound and for geometrical transformations. Software and programming languages: Max MSP and Mel script Maya. Courtesy Alisa Andrasek.

of composition by discrete units is not by itself a novelty—bricks, for example, have been around for quite a while; the novelty is that computers today can manipulate extraordinary numbers of such particles individually and each differently, if need be. These are the kinds of "customized," granular operations that mechanical machines cannot typically perform and that would be immeasurably time-consuming if carried out by hand.

As a result, designers soon started to experiment with discreteness not only as a notational tool, but also as a mode of fabrication. Discreteness then became the outward and visible sign of a new and invisible logic at play—a new post-human logic conspicuously at odds with the small-data logic of the human mind, and alien to the technical logic of all human making. No human can notate, calculate, and fabricate one billion voxels one by one. That's the kind of things computers do, and we don't. Computers can repeat a huge number of very simple operations (Cellular Automata is a good example) at unimaginable speed. As will be seen in the next section, this quantitative advantage underpins a colossal qualitative shift: due to their unprecedented processing power and speed, computers tend to solve problems very differently from the way we do.

Technologists now often use the term "Big Data" to refer to this growing performance gap between humans and electronic computers. "Big Data" means data that is too big for us, humans—but which computers can work with just fine. Some have recently also revived the old engineering bromide "Black Box" in this context—thus implying that computational problem-solving may involve something mysteriously opaque and inexplicable happening in the dark belly of computers between the queries we ask and the answers they provide. But there is nothing supernatural in computational problem-solving; the speed of electronic data processing is indeed superhuman, but only in the same sense that the weightlifting capacity of an Otis elevator may outperform that of an Olympic champion. All the same, this is where, once again, computer science has opened the door to a number of vitalist views of science and nature and offered a fertile ground for many more or less unworldly ideas common among complexists—leading in turn to an improbable resurrection of that old chimera of early computer science: Artificial Intelligence.

3.4   Degenerate Complexism
and the Second Coming of AI

Many instances of degenerate complexism in design have been recently discussed by Christina Cogdell, who remains to date the most reliable source for all inclined to complexity-bashing. Some forms of animal life have always been of interest to complexists. Swarms, in particular, have often been seen as models of self-organization: flocks of birds, schools of fish, and colonies of ants are collectives of independent agents that manage, somehow, to coordinate their behavior in the absence of hierarchical structures: each group can evidently make collective decisions but, as in a cellular automaton, each agent can only be aware of, and react to, the behavior of its immediate neighbors. John Holland suggested long ago that we should take inspiration from biology to "breed" mathematical algorithms,[58] but the fascination exerted by these mysterious forms of emergent behavior in distributed living systems is such that some designers have now taken to breeding actual animals. I know the case of a design studio which once took residence in a secluded farm, far from town, for the duration of an entire term, to learn the art and science of practical apiary. The final studio presentation featured a living bee colony, which was brought to school enclosed in a glass case (one of the bees accidentally escaped from the case, which created some commotion in the audience). I have seen design studios growing algae and slime molds, and there have been famous design pieces by cutting-edge computational designers showcasing silkworms and spiders. The popularity of the slime mold, as investigated, among others, by Claudia Pasquero, is an exception in this context, being as it is perfectly warranted in scientific terms: the slime mold is a monocellular organism that should not, in principle, possess any intelligence (as it does not even have a nervous system), yet is capable of making intelligent choices; on closer analysis, it appears that slime molds make "rational" decisions following an infralogic process that is not unlike that followed by some post-human (and post-logical) computational models of Artificial Intelligence.[59] Slime molds underperform somewhat when compared to silicon chips, which may have so far limited their appeal as actual computational tools.

Complexism has had some influence on material sciences, too, as normal phenomena of displacement and deformation of

3.4   ecoLogic Studio (Claudia Pasquero, Marco Poletto), with the Urban Morphogenesis Lab at the
Bartlett, UCL, and the Synthetic Landscape Lab at the University of Innsbruck, in collaboration
with the United Nations Development Program, DeepGreen Guatemala City, 2020. Redesign of
the municipal waste collection networks of Guatemala City using a GAN algorithm trained on the
behavior of a Physarum Polycephalum slime mold. Courtesy Claudia Pasquero.

inorganic materials under pressure—of the kind normally dealt with in structural engineering—are now often interpreted as signs of self-organization. But that is the case only if we interpret "emergence" in general as a phenomenon of vitalist animation, or if we bestow some degree of independent agency to inert matter. A BIC ballpoint pen accidentally falling off my desk is not self-organizing; it is likely just falling off my desk. Now, if the same pen would rise from my desk, nod at me, and go out for a drink with a Ticonderoga pencil, that would be a different story altogether—but one so far rarely reported.

More momentously, and pervasively, the growing reliability and almost miraculous performance of today's automated problem-solving systems has led many to celebrate a second coming of Artificial Intelligence, and this resurrection is often seen, in turn, as a vindication of some of the tenets of complexity theory. But today's born-again version of Artificial Intelligence is quite different from the one that misfired half a century ago. True, what many today still call Artificial Intelligence often works, whereas most Artificial Intelligence experiments tried over the last sixty years mostly failed; all the same, the rerun of Artificial Intelligence today as a fully-fledged design tool is problematic on more than one count. As a starter, the technical logic of today's data-driven, brute-force-powered Artificial Intelligence is so different from last century's that using the same terms, theories, and notional frame of reference for both may be misleading, or just plain wrong.

From the early days of Artificial Intelligence, in the late 1950s, computer scientists have been divided between two schools of thought. One school favored the imitation of the deductive methods of the mathematical sciences, based on formalized rules for any given task or discipline (for example, the rules of grammar for a program meant to translate between languages); the other school of thought favored inductive processes, based on iterative trial and error and on various strategies meant to reduce the number of trials. The first of these two methods imitated human science; the second, human learning; both followed established patterns of Western science (induction, formalization, deduction, rationalism, or empiricism); and neither imitated the physiology of the human brain, in spite of some terminologies then adopted and still in use ("Neural Networks," for example), which deceptively suggests otherwise.

Knowledge-based systems, expert systems, or rule-based systems belong to the first school (often also called "symbolic" due to the use of notations derived from mathematics or formal logic); "learning" systems belong to the second school, also recognizable for the widespread adoption of the "neural" (for "nervous") and "connectionist" (for "nervous systems") lexical metaphors.[60]

In 1960 Marvin Minsky, one of the protagonists of the seminal Dartmouth College summer workshop of 1956, outlined the core principles of Artificial Intelligence as a "general problem-solving machine . . . based on heuristic programming and core search" (i.e., based on massive, randomized trial and error). Over time, Minsky predicts, machines will learn to limit the "search space" of randomized trials by developing actual inferential, or inductive, reasoning skills. Thanks to inductive intelligence, humans can generalize conclusions from patterns detected in recurring events, and thus predict an unlimited number of forthcoming events based on a limited number of events that have already occurred. In 1960 Minsky thought that Artificial Intelligence would, over time, acquire that unique ability of organic intelligence, thus becoming almost undistinguishable from it. But, aware of the technical limitations of the machinery at his disposal, Minsky also conceded that the successful implementation of an "inductive" mode of Artificial Intelligence was not by any way imminent, and as a result devoted most of his essay to various alternative and, so to speak, provisional shortcutting tricks, harsh but effective, aimed at limiting the number of trials to be carried out by the machine (and many of the strategies he outlined, notably gradient-based optimization, are still in use today).[61]

Minsky's subsequent and influential career in computer science was notoriously marked by his aversion to the artificial learning model developed by Frank Rosenblatt, a Cornell psychology professor. Rosenblatt had invented an electronic device, called Perceptron, which he claimed had learning skills modeled on the physiology of the human brain (not, to be noted, on the mathematical logic of human science). In 1969 Minsky partnered with an MIT colleague, Seymour Papert (also known to this day for his participation in Negroponte's One Laptop per Child project, 2005–2014) to write a book that, while aimed specifically at Rosenblatt's Perceptron, de facto vilipended the entire connectionist school, powerfully contributing to its fall from grace and near-obliteration in the decades 115 that followed.[62] But times have changed, and today the connectionist school is back in fashion; indeed most of the recent successes of Artificial Intelligence have been achieved with connectionist, not with symbolic tools. It is therefore peculiarly odd, and noteworthy, that Minsky's first seminal essay of 1960, still available as a working paper on the MIT website, would offer one of the best definitions to date of the connectionist method—the "learning" method which Minsky would eventually make his enemy of choice. In fact, the main lines and general principles of Minsky's "heuristic" machine of 1960 would still apply in full to most connectionist tools of Artificial Intelligence today—with one technical difference however, which, as mentioned above, signals in turn a major epistemological shift.

One generation after the invention of the PC and the rise of the Internet, the novelty of today's Big Data computation (or Dataism, or Brute Force Computing) is that, for the first time ever in the history of humankind, there seems to be no practical limit to the amount of data we can capture, store, and process. This is an unprecedented, almost anthropological change in the history of the human condition. From the invention of the alphabet until a few years ago, *we always needed more data than we had*; today, for the first time ever, *we seem to always have more data than we need*. Humankind has shifted, almost overnight, from ancestral data penury to a new and untested state of data affluence. One of the first techno-scientific consequences of this truly Copernican upheaval is that many traditional cultural technologies and social practices predicated on our supposedly permanent shortage of data are now, all of a sudden, unnecessary and obsolete. By converting huge amounts of experimental data into a few lines of clean, crisp mathematical script—formulas we can comprehend within our mind; notations we can work with— modern science in its entirety functioned over time as a superb, and eminently successful, data-compression technology. But Artificial Intelligence today needs less and less of that. Artificial Intelligence today can do its job—solve problems—without any need to replicate the small data logic of either human science or human learning. This is because today's data-driven, brute-force computing (i.e., today's new version of Artificial Intelligence), is by itself a new scientific method—but one made for machines, not for humans.[63]

Scientific induction, or inference, is the capacity we have to construe general statements that go beyond our recorded experience. We have this capacity because we are intelligent, and we like to understand things. But we also need this capacity because our memories and processing power are limited, so we need to make the most of the few data we can muster. However, let's assume, *per absurdum*, and to the limit, that we can now build a machine with almost unlimited, searchable data storage. *Such a machine would not need to construe general statements that go beyond its recorded experience, because its recorded experience could be almost infinite.* Consequently, this machine would have perfect predictive skills without any need for mathematical formulas, or laws of causation—in fact, without any need for what we call science. Such a machine could predict the future by simply retrieving the past. The search for a precedent could then replace all predictive science; a Universal Google Machine would replace all science, past or future. The motto of this new science would be: Don't calculate; Search. Or, to be more precise: Don't calculate: Search for a precedent, because *whatever has happened before—if it has been recorded, if it can be retrieved—will simply happen again, whenever the same conditions reoccur.*[64]

Of course here one would need a lot of small print to define what "the same conditions" means—which would bring us back to some core tropes and problems of the modern scientific method; and indeed many traditional scientific tricks and trades and shortcuts of all kinds still apply at all steps of the new computational science of Big Data and brute force computing. Conceptually, however, and ideally, this is the main difference between yesterday's AI and today's; the main reason why AI works today and in the past it didn't. This is how computers today can translate between languages—not by applying the rules of grammar, as Noam Chomsky thought not long ago, but looking for the record of pre-existing translations validated by use. This is how computers win chess matches: not by applying the rules of the game, but by searching for a suitable precedent in a universal archive of all games already played. In this instance computers can actually simulate all kinds of fake precedents on demand, playing against themselves; and these simulated precedents will be just as good, for predictive purposes, as real historical ones. Likewise, computers can solve many otherwise intractable problems by simple iterative trial and error: trial and error, which is a very time-consuming process when carried out by humans, becomes a viable strategy when performed by computers that can run one billion trials per second.   117

Moreover, computer simulated trials need not be random, or blind: they can be "optimized"—reoriented toward areas (or "solution spaces") where results appear to be more promising. When that happens computers do something akin to learning from their own mistakes. But when a solution is found that way, it will have been found by chance: there won't be any formula, function, nor law of causation or correlation to explain why that particular solution works better than any other—nor any way to tell if a better one may or may not be found. There can be no deduction in the scientific method of brute force computation, because there is no induction—nor inferential generalization, formalisms, or abstraction. The *flat epistemology* of brute force computing lives in a timeless present where each event is worth the same as any other—as all that happens or has ever happened is equally recorded and losslessly retrievable.

This is how computers today may at times predict things that modern science cannot calculate, and our mind cannot understand. In some cases, computers can already tell us what is going to happen, but they won't tell us why—because computers don't do that: prediction without causation is prediction without explanation. This is the origin and nature of the successes of today's Artificial Intelligence. Artificial Intelligence is not a spaceship; it is a time machine. It predicts the future by having unlimited access to the past—as it happened or as it could have happened. Evidently, this is a far cry from the way we function. But what's so strange in that? Every tool we devise has its own technical logic, and a purpose; when we handle a hammer, we know what we can and cannot do with it. Ditto for Artificial Intelligence—so long as we have some notion of how it works, and what it is best at.

3.5   The Limits of AI 2.0

Evidently, today's born again, data-driven and brute-force-powered Artificial Intelligence does in many cases fulfill the expectations of its first prophets more than half a century ago: Artificial Intelligence is now a pretty effective problem-solving machine (if not a general one) hence in many circumstances a viable decision-making tool. But data-driven Artificial Intelligence solves problems by iterative optimization, and problems must be quantifiable in order to be optimizable. Consequently, the field of action of data-driven Artificial

Intelligence as a design tool is by its very nature limited to tasks involving measurable phenomena and factors. Unfortunately, architectural design as a whole cannot be easily translated into numbers. Don't misunderstand me: architectural drawings have been digitized for a long time; but no one to date has found a consensual metric to assess values in architectural design. Therefore, only specific subsets of each design assignment can be optimized by Artificial Intelligence tools, insofar as they can lead to measurable outputs;[65] even within that limited ambit, most quantifiable problems will likely consist of more than one optimizable parameter, so someone, at some point, will have to prioritize one parameter over another, and make choices. Which is the same as saying that, either way, the more general, holistic process of architectural design is destined to remain, to some extent, arbitrary. This is not something we should praise or blame; it is a matter of fact, which architectural design shares with a number of arts and sciences.

Long ago, the art of rhetoric was specifically invented to persuade an audience that one argument is better than another, even when it is not or when there is no way to tell. Likewise, the arts of design deal mostly with persuasion, not with demonstration. We can easily prove that a certain solution in a competition entry will be cheaper or more robust or faster to build or more environmentally friendly than any other—and we should: this is exactly what computation can help us do. But this may not help us win that commission, because committees and clients often base their choices on subjective, non-quantifiable factors. This has been the fate of architectural design since the beginning of time, and there is no reason to assume that Artificial Intelligence will change that more than natural intelligence ever could. And I wouldn't reiterate these truisms here if I didn't see so much human intelligence being wasted, right now, in trying to put Artificial Intelligence to so many pointless, quixotic tasks—tasks that are alien to the nature of Artificial Intelligence, and which Artificial Intelligence cannot handle.

Think of a very simple design decision—stripped down, for anecdotal clarity, to its basics. A friend must build a garage in her suburban garden. The garage is already chosen and paid for—prefabricated, bought from a catalog; our friend only needs to decide where in her garden she should put it. After legal distances and other regulations are taken into account, there is still room for plenty of choices. So we can calculate and show the position of the garage that would maximize the shadow on the summer solstice, and the position relative to the driveway that would minimize snow shoveling in winter. We can use agent-based modeling to minimize the disruption brought by the new construction to the daily open-air meanderings of the family's two big cats. We can chart all these results and more and then weigh them based on our client's priorities. This is how an algorithm would choose the position of that garage. But, as we all know, when offered this optimized result, our client will most likely say—sure, thanks, but I would still like my garage to be built right here. This happens because architecture is, besides many other things, a system of signs, which we occasionally use to convey meanings; and this also happens because, on top and above or even regardless of all the above, sometimes we choose some things because we like the way they look. Many visionaries of the quantification of knowledge have tried—and some will keep trying—to measure aesthetic appeal. If beauty can be measured, the delivery of beauty can be automated. And indeed, for some very specific, targeted projects aimed at highly structured constituencies, this might not be an inconceivable task, nor an unreachable one.

3.6   Machine Learning and the Automation of Imitation

Among many applications of machine learning one in particular, where Generative Adversarial Networks (GAN) are used as image processing tools, has recently retained the attention of computational designers. The technology has been trained to recognize similarities in a corpus of images that are labeled as instantiations of the same name, or idea—for example, one million images of dogs. The machine will extrapolate some traits common to all dog-looking images in the dataset it has been fed, then use this definition of dogginess to identify any future individual it may encounter as a dog, if the newcomer looks like all dogs known to the system. More interestingly for a designer, the machine can use this ideal type of dog to create realistic images of an infinite number of non-existing dogs, all different but all similar to their models, and to one another (insofar all dogs look like dogs). Architects tend to use the technology not for processing images of dogs but for processing images of architecture;

for example, if trained with images of Mies van der Rohe buildings, the system will learn to generate images of new buildings that look like buildings by Mies van der Rohe; alternatively, the system may be asked to tweak existing images from another set of buildings, either real or fictional, to make them all look similar, somehow, to Mies van der Rohe buildings (and this contamination of visual features between two different datasets is now often called "style transfer.")[66]

This apparently ludic display of computer graphics prowess deals in fact with some core theoretical tropes of Western philosophy and art theory. Modern image theory can be traced back to a trend that started at the end of the Middle Ages, when the Aristotelian framework of Scholastic philosophy started to give way to a Neoplatonic one. In the Scholastic tradition ideas were seen as sets of predicates, and no one in the Middle Ages imagined that such predicates could be other than nominal (i.e., simply, names). Beginning with Marsilio Ficino, however, Platonic ideas came to be seen increasingly as images—visual images, living in the mind of God, and more or less corrupted in our sublunar skies, depending on the degrees of separation from their archetypes (which Renaissance Neoplatonists could calculate with excruciating precision).

Regardless of what we would call their resolution, these epistemological images were seen as visual, not verbal, constructions, hence this idea of what ideas should look like soon started to merge with the then up-and-rising discipline of art theory. Famously, with Bellori (1664) Neoplatonic ideas became one and the same as the "Ideas of Painters, Sculptors, and Architects," a fusion of philosophy and art which had been in the making since the early Renaissance, and which served Bellori's classicist project well: just like Neoplatonist philosophers believed in visual ideas as the abstract matrix common to all worldly manifestations, an artist's idea, in Bellori's theory, should show nature's general intentions, not any particular instance thereof: painters should represent things not as they are but as they should be.[67]

As to how artists would manage to pierce the carapace of mundane appearances to capture or intuit a glimpse of the timeless ideal hidden beyond them, opinions diverged. Raphael famously claimed that, to avoid copying reality too closely, he followed "ideas" that "just sprang to his mind."[68] Others tried to be more specific. But that was—and still is—a tall order, because the clash between 121 realism and idealism in the mimetic arts goes straight to the heart of the classical, then classicist, agenda: if both nature and art are copies of otherworldly ideas (never mind in which order, or who's copying whom), and Platonic ideas are themselves visual images, how do we define a successful act of creative imitation in the visual arts?

Starting from classical antiquity, many concluded that the best way for an artist to overcome and transcend the identical, mirror-like replication of a natural model is to imitate not one but a set of chosen models, to be inspired by all and none in particular, taking the best from each. But, as will be seen in more detail in the next chapter, that opened a Pandora's box of theoretical and practical issues. How do we put together a collection of suitable models, which must be all different but also compatibly similar, so we can fuse and merge them all, somehow, in a single act of creative imitation? More in general, but crucially, what do two images (or bodies or buildings) have in common when they *look similar*? What do a mother and her biological daughter have in common, for example, if they only vaguely resemble each other? Today we know that they share some genetic code—but that's only a recent discovery, and besides, genetic code is invisible. Herein lies the secret and mystery of creative imitation: how do we define resemblances and similarity in the visual arts, in literary compositions, and more generally in visual or aural perception? Scholars and artists of classical antiquity and early modernity, from Cicero and Pliny to Bembo, Erasmus, Bellori, and Quatremère de Quincy, and then in more recent times scientists and psychologists have tried to answer that query.[69] Today the question may no longer apply, because machine learning (in this instance, GAN) has successfully started to *automate imitation*.

Thanks to data-driven Artificial Intelligence we can now mass-produce endless non-identical copies of any given set of archetypes or models. The GAN technique can produce similarities working the analytic way—by abstracting one ideal archetype out of and common to many similar images—or, by reversing the process, generating many realistic images similar to their models; by applying the two processes sequentially, GAN can produce a new set of non-identical copies collectively similar to one or more original datasets. Either way, it would appear that Artificial Intelligence has finally cracked the mystery of creative imitation, which had baffled or eluded art theory and the theory of knowledge since the beginning of time. And in true Artificial Intelligence fashion, we can now use machines to produce imitative (i.e., non-identical) copies without knowing what imitation is—and without knowing how it works: in this, we are not an iota more advanced than any of our predecessors.

As discussed at length in the first chapter, the mass production of variations has been a core principle of parametric fabrication from the very start; GAN have now extended the same technical logic to include some very specific sets of parameters—those determining the visual similarity, or resemblance, between an original and its non-identical copies. Among the most alert interpreters of GAN technologies, Matias del Campo and Sandra Manninger have recently pointed out that a GAN machine learning algorithm, when trained on a consistent dataset of artwork, does something akin to *learning the style of the dataset it has been fed*; and when used generatively, does something similar to producing new artwork *in the same style*. This computational definition of "style" as the tacit common denominator of a visually consistent dataset is not more arbitrary—and it is certainly more helpful—than many that have been offered by scholars over time.[70] And it is meaningful that today's machine learning should foster a re-evaluation of the role of imitation and style in the arts of design, many decades after an unstated modernist mandate more or less expunged both terms from architectural discourse.

In practice, and in real life, architectural imitations, including the stealthy imitation of tacit architectural "styles," are as pervasive and ubiquitous today as they have always been. Yet, for better or worse, creative or stylistic imitation are not relevant topics in the arts of design these days, nor have they been for more than a century; imitation, which in the classical tradition was seen as an essential component of every creative endeavor, was demoted and jettisoned by late-romantic artists and then, for different reasons, by modernists, who also felt, somehow, that all notions dealing with visual similarities or resemblances led to intractable, non-quantifiable design issues. As a result, imitation in general is still often seen today as a form of plagiarism.

This may now be about to change. Once again, advanced computation and Artificial Intelligence are prompting us to reassess some aspect of our natural intelligence that twentieth-century industrial modernism made us forget; if we are lucky, soon we may 123

3.5, 3.6   SPAN (Matias del Campo, Sandra Manninger), Generali Center Mariahilferstraße,
proposal, Vienna, Austria, 2021. A dataset of 7,860 images of Brutalist buildings was automatically
scraped from the web, then processed via a StyleGAN2 to create a "latent walk" through a landscape
of images that are strange but familiar enough to be perceived as similar to existent buildings.

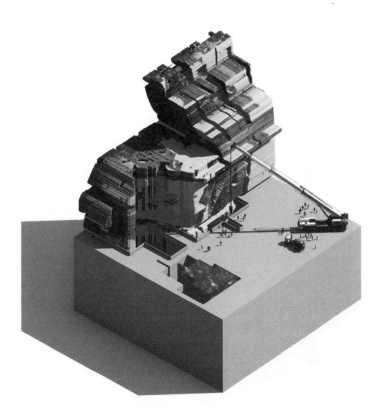

The results of the data scraping were in turn the starting point for a project consisting of a shopping arcade, cineplex, and offices. Using a neural network that allows the pixels of the images to project onto a mesh, a selection of images was converted into 3D models for further project development. Courtesy Matias del Campo.

also reacquire some creative and critical capacity to deal with similarities, resemblances, styles, and imitation in the arts—just as we always did; and now having at long last at our disposal a technology that actually seconds that. Yet, at the time of this writing, the relevance of GAN technologies to contemporary architectural design still appears as somewhat limited: which designer would want to borrow someone else's intelligence (never mind if artificial) to design a building that looks like someone else's building? A similarly cautionary note should also apply, in my opinion, to the more general issue of the practical applicability of data-driven Artificial Intelligence to architectural design.

3.7   Sorry: There Won't Be a Third Digital Turn Driven by AI

As mentioned, computational problem-solving can only solve quantifiable problems, and some crucial design choices are not quantifiable—even though many subsequent, ancillary decisions are, and this is where Artificial Intelligence will be a game changer, and to some extent it already is. But many current efforts to use Artificial Intelligence tools outside of these technical limitations are ultracrepidarian (as the painter Apelles would have said), and wasteful. Crucially, when we do use AI tools (and all suggests we shall use more and more of them, and more and more often) we should be mindful of the fundamental difference between the way computers think (or, strictly speaking, solve problems) and the way we do. We now know that the best way to exploit the immense powers of AI is to let computers solve problems in their own way, without imitating the processes of either human science or human learning; Artificial Intelligence is a new, fully fledged scientific method—but a post-human one, made to measure for electronic computers. We cannot solve problems that way—anymore than computers can replicate our way of thinking. And we should beware of unduly generalizing the technical logic of computation, by carrying it over to where it does not belong.

Yes, computers are post-human, intelligent tools: good for them—not that they care—and good for us if we can use AI to solve problems faster, or to solve problems we could not solve in any other way. But that doesn't mean that the methods of computation—the technical logic of how computers work—would necessarily help us understand the inner workings of a school of fish, of the financial markets, of a metropolis, or of the human brain. The postmodern sciences of complexity, emergence, and self-organization, now so deeply intertwined with computer sciences, have proven remarkably effective in describing the operations of all things computational—from the internet to neural network backpropagation. In fact, one of the reasons why complexism is so widespread today is that it has proven, oddly, more successful in describing technical networks than it ever was in explaining natural phenomena, where the record of complexity science as a whole was always patchy. But the extension of complexism from inorganic, biological, and technical systems to social sciences, and to society as a whole is not a "new kind of science"—it is a political ideology: and in my opinion, one of the most pernicious and nefarious ideologies of our time.[71] True, the internet can self-organize. That's neither magic nor destiny, and it was never inevitable: we designed it that way. Apparently, it was an excellent idea. But if we let pandemics self-organize we let millions die; if we let societies self-organize, we relinquish democratic life.

4   The Post-Human Chunkiness of Computational Automation

Humans cannot easily manage large stacks of unsorted items—think of millions of different items randomly dumped and heaped in a warehouse.[1] The usual way for us to deal with a mess is to give it some order; for example, to find a name among one million we typically invest a lot of upfront work to sort all those names alphabetically; that investment pays back each time we look for a name, because then we know in advance where it is, and we don't have to read one million names to find the one we are looking for. The same in mathematics: to handle one million mathematical points we typically inscribe them in an equation, so we deal with just a few lines of algebraic script instead of having to deal with one million coordinates. In applied sciences, as in design, we do not handle dimensionless mathematical points, but physical particles, pixels, or voxels—chunks of images and of three-dimensional objects; yet the logic is the same. No human can notate and calculate one billion parts (chunks, pixels, voxels) one by one. Life is too short for that. When dealing with data, simplification is the humans' inescapable lot. That's the way our mind works.

In the past, pre-industrial artisans could sometimes produce artifacts of extraordinary, "irrational," extravagant complexity precisely because that complexity was never datafied—it was neither calculated nor notated; in short, it was not designed. It was just made, and sometimes, when the master builders were inspired, or got lucky, all the parts they put together kept together, and the building stood up. But when modern science stepped in, and we started to make things by design and by notation, data simplification became the norm—because there is only so much data that humans can take in and process at any given time. Slowly but surely, we are now realizing that no such limits apply to the post-human science of computation. Computers can deal with the most overwhelming, unintelligible, meaningless mess—think of the typical Amazon warehouse—due to their sheer speed, memory, and processing power. Quite unlike us, to find a name among many in a list a computer would indeed read them all. Computers can manage an almost never-ending raw list of unsorted mathematical coordinates just fine.

As discussed at length throughout this book, the anti-human, irrational messiness that computers can typically get on well with, and navigate at ease, is already the index of a post-human intelligence at play; of a technical logic that is no longer the logic of our mind. More recently, with the rise of intelligent robotics—both of the notational and of the post-notational kind—this technical logic has spread from abstract notations and calculations to logistics and fulfillment, then to actual material fabrication, and what used to be a post-human way of thinking has grown to be a new, fully fledged and equally post-human way of working.

What many now call "the second digital turn" started to unfold when a new generation of digitally alert designers decided to endorse, embrace, and display the technical logic of the new computational tools they used. Early in the second decade of the new century, discreteness, roughness and chunkiness replaced the continuous lines and surfaces that had marked the style of the first digital age, and "excessive resolution" became the visual marker of the new age of data-driven, brute-force-powered computation. Just like "big data" means data that are too big for us, but not for computers, "excessive resolution" means the display of a level of granularity that computers and computer-driven fabrication tools can work with just fine, but which is beyond reach for traditional hand-making or for mechanical production.[2]

For all that, discreteness and chunkiness were—and are—design choices. Computers and robots are versatile machines; as all tools, they feed back onto and influence the way we work, but at the end of the day we set their agenda, not the other way around. Some second-generation digital designers may have embraced discreteness as a visceral, almost knee-jerk reaction against the overflow of digital streamlining that marked the first age of computational design, and against the spliny style that was dominant in the digital studios where they taught and trained. But in so far as "splininess," or spliny smoothness, was a deep-rooted theoretical trope of the first digital turn, its rejection was bound to acquire equally deep theoretical implications, far outstripping the usual boundaries of generational rebellion.

Indeed, with a few years' hindsight, it already appears that the opposition between continuity and discreteness that has animated the last wave of digital design, and which in the case of computation was nominally borrowed from mathematics, may just be the latest avatar to date of the opposition that art historians always saw between the smooth and the rough; that technologists used to see between the assemblage and the fusion; that classical literary critics saw between citation and influence, and post-structuralist literary critics between text and intertextuality; that early digital artists used to see between collage and morphing, and architectural theoreticians of late modernism saw between the whole and the part. It is the timeless opposition between the block and the aggregate, unity and addition, matter and particles, the monolith and the heteroclite—between *things that are made of one piece, and things that are made of many.*

4.1 Mechanical Assembly as the Style of Dissent

We can make things by taking a chunk of solid matter, as found, then removing matter from it as needed, until we get to the shape we want; or we can proceed the opposite way, by picking and choosing a number of smaller chunks, either found or made, and adding them to one another, somehow, until we get to a bigger chunk. When we make stuff the additive way, the smaller chunks, parts, or ingredients we mix and match may either show in the final product, or they may merge in a single, uniform block. In the latter case, we end up with a monolith—although an artificial, not a natural one in this instance. In the former case, we have a heteroclite (in the etymological, not in the current meaning of the term): a whole that is made of discernible, discrete, or separate parts.

Classical antiquity saw art as an imitation of nature—in the simplest, most literal sense: a painted apple had to look as much as possible like a real one; Apelles was famous for once painting a horse so perfect that its image fooled even other horses. As nature often produces monoliths (example: a big boulder), while purposeful heteroclites are generally the result, hence the sign, of human laboriousness (example: a drystone wall), classical artists aimed at merging the parts of their composition in a single, uniform, smooth, and homogeneous whole—as smooth as if made by nature itself. This principle is implied in one of the most influential *topoi* (anecdotes, legends, or parables) of classical art—one of those apparently dumb little stories that the Greeks and Romans often used in lieu of a fully fledged theory of what we now call the visual arts. This is the story, as told, among others, by Cicero and Pliny: the famed Zeuxis, the best-paid painter of his time, was invited to a town in what would now be southern Italy to paint a picture of a goddess. In search of inspiration he asked to see some examples of local beauties. The town elders sent him a selected group of handsome young men. Zeuxis protested, and he was then allowed to see some girls, but finding none quite to his taste, he retained five of them as models. His painting merged features taken from all five, and it met with great success—hence the lasting popularity of the anecdote. From the point of view of art theory however, and even of the theory of human knowledge, that seemingly innocent tale conceals a number of major theoretical conundrums.[3]

If Zeuxis already had an idea of feminine beauty in his mind, why did he need to imitate any real-life model? And if, on the contrary, he did not have an innate idea of beauty, how could he choose among so many incomplete manifestations of the ideal? The conflict between realism and idealism has found different solutions throughout the ages, but from a more practical point of view, the technicalities of Zeuxis's mode of artistic operation—the parsing, selection, and the reassembly of parts coming from many models—have equally invited and prompted a never-ending stream of theories and speculations. Evidently, the artist would not have limited himself in that instance to just cutting and pasting a number of pieces, as in a jigsaw puzzle; he would most likely have had to rework, modify, and adapt some of the parts, not only to chip off the edges but also more generally to make them blend with one another and merge in a single, harmonious composition. Hence the question: how much of Zeuxis's operation was what we today could call a collage, and how much of it would have been some looser form of imitation—the work of a talented artist only vaguely and distantly inspired by some of his models, or sources? Could one look at his finished painting and tell: see, these are Emily's eyes, Peggy's nose, and Nancy's lips? Or did the artist blend all of his sources in one transfigured, truly supernatural composition, where one would say: see, there is *a certain undefinable something* in this portrait that reminds me of Emily, and of Peggy, and of Nancy, but it's hard to tell what, precisely, comes from each? In classical art theory, this is where art would have equaled nature, because this is the way nature works: this is the way a daughter looks similar to her biological mother.[4]

The opposition between these two modes of composition— between making by way of the merger and fusion of parts into a whole, and making by way of addition and assemblage of parts that 133 remain separate—was clearly outlined in classical art theory from the very start, but due to their noted logocentrism the Greek and Romans devoted most of their critical attention, as we would say today, to the way texts are made—not images. In classical and modern rhetoric the good writer is expected to learn the art of writing by reading and re-reading the best authors and becoming intimately conversant with their style, capturing their quintessence to assimilate them all, then transcend the style of each to achieve a new and superior, unique and personal creation. The bad writer, conversely, will stick too closely to his sources; a stupid, uninspired imitator, he will unduly cut and paste words, idioms, turns of phrases, and entire expressions from each model and make a mosaic or a patchwork at best, which in ancient rhetoric was called a cento. As all classical art was imitative, imitation did not have back then any of the negative connotations the term has acquired in more recent times; in the classical tradition all the difference was between the good and the bad imitator. The good imitator—the "creative" imitator—copies from many models like bees make honey: culling the fruit of many flowers. Like a honey-making bee, the good imitator keeps all the flavor of his sources, but none of his sources can be clearly discerned, traced, or tracked in the final product of his creative endeavors. The bad imitator, whom the ancients also called a "monkey," mindlessly and mechanically copies and repeats snippets picked up here and there.

All these *topoi* in defense of the creative value of transfigural imitation flared up, after a centuries long slumber, during the Renaissance, when the imitation of classical sources became a vital concern for several generations of humanists: strange as it may appear to us today, around the time Columbus landed in America a humanist was first and foremost a scholar specializing in the imitation of Cicero's writing (which was then seen as the best of all possible languages, never mind that Cicero wrote in a language that nobody spoke anymore). At the same time, however, the practice of cut-and-paste imitation unexpectedly acquired a new role and unprecedented importance, due to a drastic socio-technical change in the way texts themselves were being copied and reproduced.

Books, always rare and expensive, were all of a sudden made cheap and ubiquitous by print. Access to scribal copies of even canonical texts used to be so impervious that early humanists often ended up learning pages and pages of Cicero by heart—not so much out of choice as out of necessity. Soon, scholars could keep printed copies of almost any Cicero they needed handy on their desktops, to consult and use at will. Printed indexes, thesauri, and alphabetical or thematic sorting made this content searchable—in unprecedented ways. As a result, citing and quoting, cutting and pasting (sometimes literally) from such easily available sources became pervasive and even fashionable. Early in the sixteenth century the elegant humanist Erasmus even wrote a perfidious—and, at the time, hugely popular—pamphlet against the new fad of cut-and-paste imitation. He called the cut-and-paste imitator "the sick guy" (Nosoponus), and he ridiculed him as an uninspired, cretinous fool.[5] To no avail: the practice of cut-and-paste imitation kept spreading, and soon some even started to theorize it. Prominent among these, the Neoplatonist polymath Giulio Camillo (also known at the time for his talents as a lion tamer) built and marketed his mysterious Memory Theater as a semi-automatic device for the composition of new texts in the most accomplished Ciceronian style—on all kinds of subjects, including some on which Cicero never said a thing.[6]

Not without a certain amount of matter-of-factness, Camillo appears to have concluded that at the end of the day even the most ineffable style of writing must be, and cannot be but the result of plain and mechanical assemblages of words—words, sequences of words, expressions, and phrases; by breaking down all Cicero's writings into lemmas of all sorts and sizes, then sorting these entries to facilitate their retrieval, Camillo's Theater would have allowed any writer to compose any new text by simply decomposing, then reassembling, a suitable choice of Cicero's words and phrases in a new context. In Camillo's theory every text is a collage of older textual chunks: in the case of his Ciceronian machine all chunks were carefully and exclusively excerpted from the corpus of Cicero's original writings, but Camillo's method of cut-and-paste composition was so general and widely applicable that it could have been easily extended to other contexts and contents.

And it was. An early visionary of the mechanization of knowledge—and of the arts—Camillo had some interest in architectural theory; he was also a friend of Sebastiano Serlio. As he explained in a long lost manuscript (found and published only in 1983),[7] the analogy between the architect's job and the writer's is self-evident: humanist writers, like him, wanted to reuse Cicero's Latin to express new ideas; humanist architects, like Serlio, wanted to reuse the forms of classical architecture to build new buildings; therefore, the program being the same, architects and writers could use the same rules to play the same game. When translated into architectural terms, Camillo's cut-and-paste, or citationist method, called for: establishing a corpus of chosen monuments of classical antiquity; breaking them up into smaller chunks, cut into sizes fit for combining with others in new compositions; last, compiling and publishing a catalog of ready-made parts complete with the instructions needed for their assembly—as in an IKEA kit today. This is what Serlio did, starting in 1537.[8]

Serlio's multivolume treatise in print did not contain actual spolia from old buildings, but it featured long lists of virtual modular chunks, duly sorted according to a sophisticated arborescent hierarchy, each drawn in plan, elevation, and section (or almost), together with sets of rules for the redesign of each chunk at different scales and in different contexts as needed. These rules did allow for some leeway and occasional alterations in the design of each part, but a drawing (and a building) made by the montage of many chunks is bound to look chunky.

Serlio unapologetically acknowledged the ungainly, clumsy chunkiness of his designs, and he was evidently aware of the unusual, nonconformist spirit and ambition of his program. For his was a deliberately simplified design method, meant to teach the basics of architecture to all and sundry; meant to turn design into a professional, run-of-the-mill technical operation: an operation that almost anyone could easily learn—and, crucially, learn from books. Not surprisingly, Serlio never had good press, and long remained unpopular among design historians;[9] only in more recent times Tafuri and others have related Serlio's oddly pre-mechanical style to his Evangelical, crypto-Protestant, possibly Nicodemite proclivities.[10] Some of the same chunkiness, indeed at times even more extreme, in Michelangelo's late architectural works was, and still is, less harshly received. But through Serlio's work, and partly as a side effect of his notoriety as a philistine *avant la lettre*, architectural chunkiness earned an early and equally unpalatable, albeit at the start mostly subliminal, reputation as the sign of a utilitarian, unrefined, materialistic, and pre-mechanistic view of the world—the view of a modern world that was then barely dawning.

4.1   Sebastiano Serlio, *Livre Extraordinaire de Architecture* [...]
(Lyon: Jean de Tournes, 1551), plate 18.

4.2   Michelangelo, Porta Pia, Rome (inner face, begun 1561).
Photo courtesy Antonino Saggio, Rome.

From today's vantage point, it is not surprising that the early modern rise of mechanical reproducibility in print may have favored a method of composition based on the assembly of ready-made, mechanically reproducible fragments or citations—both verbal (via ready-made chunks of texts) and visual (via ready-made chunks of images). As discussed in the first chapter, print was the first technology of mechanical mass production where the replication of identical, standardized parts was meant to generate measurable economies of scale—and this was not limited to the reproduction of alphabetical type: woodblocks for printing images were sometimes disassembled into tessellae that could be inserted, mosaic-like, into other woodblocks, to repeat some standard visual motifs or even less standard ones, such as heads or figures or landscapes, at a lesser cost. Such composite woodblocks were in fact modern collages; cut-and-paste images in all but name.[11]

Unlike "the divine" Michelangelo, a lifelong rebel and troublemaker who nonetheless kept working for the Popes to his last day on earth, Serlio died poor, angry, and forlorn in Lyon—a religious refugee of sorts, apparently earning his living as a woodcut and metal plate engraver for a Huguenot printer of Calvinist tracts and Geneva bibles.[12] But his revolutionary treatise, noted for his bare-boned version of the classical orders and for an equally iconoclastic, perfunctory approach to the design of standardized building types—including, oddly, housing for the poor, or social housing, as we would say today—was long influential in Northern Europe and in Protestant countries; through reprints and translations, and also via cheap popular pattern-books largely derived from it, it inspired a style of Nordic, classicist chunkiness that John Summerson called "Artisan Mannerism."[13]

While the Catholic South embraced the curvy grandeur and aristocratic smoothness of counter-reformation baroque, the Protestant North often kept showing a distinct preference for a plain style of visual parataxis—at least, this is what disjunctions would have been called by rhetoricians back then, if they had been looking at visual languages the way we do now: the southern baroque likes to smooth the transition between two flat or angular surfaces with a profusion of sinuous moldings; northern classicism is not afraid of stark, unmediated juxtapositions—the former mode of composition favoring the whole, the latter the parts. In countries torn by religious 139 conflict, as England was between Commonwealth and Restoration, it would be tempting to see the choice of baroque sinuosity as a sign of Papist allegiance, and classicist disjunctions—or discreteness—as a Puritan alternative to the Roman model; Hawksmoor's at times obnoxious chunkiness has recently been seen as a form of precocious Brutalism.[14] These generalizations may of course be far too sweeping and unwarranted, but given the Serlian precedent and legacy, it is nonetheless arguable that classicist chunkiness may have been early on associated with ideas of pedestrian utilitarianism, anti-artsy plainness, and no-nonsense, Puritan sobriety, thus suggesting from the start a somewhat oppositional, even revolutionary stance in religion and politics. If so, architectural chunkiness may have been seen and felt by many—hard to say since when—as an expression of dissent.

4.2   Modernist Modularity, Postmodernist Collage, and Deconstructivist Aggregation

Be that as it may, the mechanical connotations of the cut-and-paste way of making became inescapable when, a few centuries later, modernist art started to tackle, critique, and question the technical logic of industrial mass production. Braque's, Picasso's, and Gris's collages were assemblages of printed (or stenciled) typographical characters, of mechanically printed wallpaper, or of actual newspaper pages; Duchamp's ready-mades were montages of standard items of mass production: bicycle wheels, snow shovels, bottle racks. Throughout the twentieth century assemblage, montage, collage, cut-and-paste, and visual citations, albeit primarily executed by hand, visually displayed the manipulation of mass-produced, industrial items; the very notion of assemblage in artistic production indexed the industrial assembly line, and assemblages were used as an artistic device to represent and critique the industrial, mechanical way of making. Cut-and-pasted chunkiness in the arts stood for industrial modernity, because industrial mass production and assembly-line modularity tend to generate chunky stuff: each fabricated part can of course be as streamlined as needed, but the tectonic logic of mechanical assembly still requires that separate chunks, no matter how smooth, be joined together in a heteroclitic, hence chunky, whole.

140   In 1959 Clement Greenberg, the pontiff of high modernism, wrote

4.3   Nicholas Hawksmoor, Christ Church Spitalfields, London (begun 1714).
Photo courtesy Owen Hopkins, London.

a famous essay, still curricular reading in departments of art history around the world, extolling collage as the core theoretical trope of modernist art.[15] The formalist Greenberg would have been loath to admit it, but his arguments would have equally applied to the modes of production that collage stood for, portrayed, and referred to.

Fast forward to the end of modernity—and to the postmodern turn of the late 1970s. Given the state of affairs I just described, collage and citation should have been, back then, unlikely candidates for PoMo adoption. If you are a PoMo militant in, say, 1978, why would you care for a set of stylistic signifiers and compositional devices that were then universally seen as staples and icons of modernity—that is, of your own chosen enemy? Yet, as we now know, PoMo designers of all ilk soon embraced citationist chunkiness without any reservations, apparently oblivious to the modernist lineage and credentials of mechanical collaging, gluing, and pasting. This unexpected development was mostly due to one book and to the influence of one of its two authors. Still a student (actually, it appears, Rudolf Wittkower's only master's student at the Warburg Institute in London) Colin Rowe had begun his career by claiming that, some four centuries apart, Palladio and Le Corbusier were in fact up to the same thing—one in plan the other in elevation.[16] One generation later the anti-modernist crusade of Colin Rowe scored another home run.

The core argument of Rowe's *Collage City* (1978, with Fred Koetter) was that architects had been led astray by their faith in technology and science or, in other cases, by their subservience to popular taste. As an antidote to both fallacies Rowe suggests that architects should re-establish "a sceptical distance from big visions of social deliverance,"[17] i.e., that architects should abandon all hope and ambition of doing something good for the rest of the world. As practical means to that end, Rowe offered two models: collision city, the city of bricolage, of which the archetype is Hadrian's Villa in Tivoli; and collage city, the city as a museum, of which Rowe's best example was Biedermeier or Restoration Munich (the historicist city of von Klenze and von Gärtner); Rowe's second choice was, curiously, the little town of Novara in Piedmont. Both collision city and collage city, as their names suggest, are assemblages of disjointed fragments, and are seen as the result of relatively uncoordinated additive processes, in the absence of a unified urban design; in the case of collage city, the fragments are citations—and in the case of Munich, verbatim replicas—of monuments from different periods of architectural history, all of them referring to the classical tradition (and it may as well be that British-born Colin Rowe, 1920–1999, did not know or acknowledge any other architectural tradition outside of that one).

So there you go: in a few pages, thanks to Rowe's droning and often vapid, wordy prose, Braque and Picasso's modernist collage was turned into a formal game of linguistic reference to the history of European classicism; and chunkiness and disjointedness—often the sign of the belligerent or antagonistic stances of various activist avant-gardes throughout the twentieth century—became the *signe identitaire*, and the rallying cry, of a new wave of architects whose main project was to have none. To be noted, Rowe was not alone in plotting that chart back then, and in many ways his collage urbanism was very much in the spirit of the time: the itinerary of the architectural collage, from modernist modularity to historicist citation, is parallel and similar to the coeval drift of the critical notion of intertextuality, born as a technical offshoot of modern structural linguistics with Julia Kristeva and Roland Barthes in the late 1960s and early 1970s,[18] but which would soon thereafter come to mean something akin to a theory of endless textual referentiality—a game of mirrors whereby, via citations or allusions, a text deliberately refers to another, and then to another, *ad libitum* and possibly *ad infinitum*, to the detriment of denotative meaning: if "every decoding is a new encoding," all interpretation is worthless, and all communication impossible.[19] So for example Jean-Luc Godard's references to the history of cinema and to film theory in his *nouvelle vague* masterpieces of the early 1960s were meant to hone cinema's power as a militant art of the index; Quentin Tarantino's game of citations in *Pulp Fiction*, to quote a famous line from the movie itself, "doesn't mean a thing" (Butch's actual line to Esmeralda Villa Lobos in episode 5 of the movie cannot be quoted verbatim). *Pulp Fiction* is a postmodern movie: it has no message at all. It certainly does not convey any urgent "vision of social deliverance."

So we see that, as the postmodern wave unfolded, a third set of meanings came to be associated with the "chunky signifier" in the visual arts: chunkiness, formerly the style of mechanical assemblage, as well as the chosen style for dissenters of all sorts—the generic style

4.4   Peter Eisenman Architects, Aronoff Center for Design and Art, Cincinnati, Ohio, 1988–1966.
Courtesy Peter Eisenman. Photo © J. Miles Wolf.

of dissent, so to speak—ended up also being chosen by some, against all odds, as the style of "no sense"; the style of those who choose to speak to deny all worth to words. And to make things even murkier, after its PoMo adoption in the late 1970s chunkiness—or at least the style of assemblage, addition, and aggregation—was also adopted, some ten years later, by some of the most virulent anti-PoMo Deconstructivists. That was at least a more predictable development: the rough assemblages and disassemblages of part, so frequent in Deconstructivist architecture, evidently evoked or suggested, at least visually, some of the trappings and trimmings of a machine-made environment. Moreover, architectural Deconstructivism was an "oppositional" movement from the start, in the tradition of the historical modernist avant-gardes of the early twentieth century, with a strong penchant for dissent and protest; last but not least, one of its protagonists and leading theoreticians, Peter Eisenman, adopted the term "aggregation" almost from the start, and made it one of his Deconstructivist battle cries.[20] However, definition in this instance didn't immediately follow proclamation, and it appears that Eisenman waited for more than thirty years before offering a retroactive conceptualization of one of the key words for his lifelong body of work. This happened only recently, in the context of a series of articles commissioned by the architectural journal *Flat Out*.[21]

In the first of those articles Peter Eisenman implicitly compares the looks of his early Deconstructivist assemblages to the messy chunkiness that has recently become the visual hallmark of some of the young designers of the second digital turn (Gilles Retsin, and Daniel Koehler in particular, who must be credited for bringing to the fore the compositional logic of digital mereology).[22] Yes, their digital work, and the abstract complexity of vintage architectural Deconstructivism may have something in common, Eisenman admits; yet their meanings are different. The "difficult forms"[23] of Deconstructivism, circa 1988, were meant to emphasize our rejection of the conventional, technocratic way of building of those times: Deconstructivism was difficult to look at because it was meant to be difficult to build. That, Eisenman argues, was our way of going counter to the architectural Zeitgest of our time—of expressing our "resistance to power." To the contrary, the visual messiness of today's digital design is very easy to build. That's just what today's robots and computers do; that's the way they work. So, the aggregational style of the second digital turn is a way of going *with* the Zeitgeist—not against it.

The argument is sound, and it could be generalized. Think of Mendelsohn's Einstein Tower, or Kiesler's Endless House: their organic forms were meant to be almost impossible to build when their designers first conceived them. Back then, that was a way for Mendelsohn and Kiesler to spell out their rejection of the dominant technologies of their time—that was their way of saying "no" to the technical logic of mechanical mass production (or, more specifically, to the technical logic of post-and-lintel reinforced concrete or steel construction). But when similar organic shapes were adopted by Greg Lynn, for example, or other blob-makers of the 1990s, that was a way for them to display their enthusiastic endorsement of the digital technologies of *their* time—that was their way of saying "yes" to the technical logic of digital mass customization (or, more specifically, to the technical logic of digital spline modeling). Very similar organic forms were bearers of opposite meanings at different times.

In the same article however, alongside this technical and tectonic view of architectural semantics, Eisenman also seems to allow for another and far more general mode of architectural signification—one that is often taken for granted, at least in common parlance and in daily life: some smooth, curvy, seamless forms would appear to be, somehow, endowed with an almost irrational power to convey feelings of compliancy and forthcomingness; whereas rough, angular, and discrete forms would be equally inherently and generically capable of carrying oppositional, antagonistic, or abrasive statements. Yet behind this apparently anodyne truism lies an epistemological can of worms. To what extent can we really posit that some forms—or sounds, for that matter—may carry a universal, almost bodily and physical meaning to all humans, past present and future, regardless of place, culture, and civilization? Philosophers and psychologists have been arguing this matter for quite some time—in fact, forever; and to date, inconclusively.

In a famous experiment first published in 1929 the psychologist Wolfgang Köhler compared two abstract visual shapes, one rounded and one angular, and two made-up names, "takete" and "maluma." When asked, most subjects associated the name "takete" with the angular shape, and "maluma" with the round one—apparently, regardless of language, culture, or environment.[24] This famous experiment was eventually used to corroborate various disputable theories. Should we take some additional steps down that slippery slope, and assume that taketian shapes may have some cross-cultural power to hurt, or irk, whomever happens to behold them, and malumian shapes an equally timeless capacity to soothe, please, or heal? In that case, we might also assume that some peculiarly chunky shapes and forms may have an equal universal capacity to offend their viewers, just like a punch on the nose or a slap in the face delivers an immediate message to whomever happens to be on the receiving end of any such unsavory medium. Art historians have been using plenty of empathic, emotive, or bodily metaphors since the rise of art history as a discipline, thus tacitly admitting, and sometimes asserting, that some visual signs may speak to the senses without going through the mediation of reason and culture. In architecture, a similarly sensual approach to visual communication has been traced as far back as Vitruvius.[25]

For our present purposes, however, there is no need to delve so deep. The story we have been recounting proves that, whatever the cause—natural or nurtural—disjointedness and aggregation in things that are made of many parts tend to be shown, and seen, as the indexical trace of some unwarranted or unresolved effort in the labor of making, hence as a sign of distress, displeasure, fatigue, or dissent. In short, they hurt. There are many historical reasons for that, and we have seen that the additive way of making did occasionally convey other sets of meanings at different times in the past—and it still does. This may create some confusion in current design criticism, as at the time of this writing (summer 2021) chunkiness and collages (particularly digital collages, i.e., fake collages simulated via computer-generated images) appear to be equally fashionable and are being invoked by all and sundry from the most unexpected quarters and for a range of disparate reasons.[26] However, in this instance the writings and statements of many of today's digital designers may help us separate the wheat from the chaff. After all, when designers spell out what they have in mind, we might as well take their words at face value. And what they say—*viva voce*, even more than in print—is evidence of a general state of mind that is quite unlike that of their immediate predecessors.

Digital designers of the 1990s actually loved their computers. Back then, personal computers in particular—the first computers 147

4.5   OFFICE Kersten Geers David Van Severen, Water Tower, competition design, Beersel, Belgium, 2007–2008. Design team Kersten Geers, David Van Severen, Steven Bosmans, Michael Langeder, Jan Lenaerts. In collaboration with UTIL Struktuurstudies. Courtesy OFFICE Kersten Geers David Van Severen.

4.6   Gilles Retsin Architecture, installation at the Royal Academy of Arts, London, 2019. An assembly of discrete timber building blocks, meant as a spatial fragment in real size of a larger building. Courtesy Gilles Retsin. Photo: NAARO.

4.7 Gilles Retsin Architecture, Diamonds House. Proposal, 2016. Project for a multifamily home
in Belgium consisting of modular timber elements. The elements have three different scales, and
are robotically prefabricated from standard timber sheet materials. Courtesy Gilles Retsin.

4.8   Daniel Koehler, University of Texas at Austin, The Three-Nine, proposal, 2022. Three
architecture machines arranged 382,659 building parts into variations on a nine-square grid
simulating the daily spatial uses of 500 people in downtown Austin, TX. Axonometrics.
Courtesy Daniel Koehler.

people could own and use in daily life—were appealing, desirable, good looking, and pleasant to the touch. Not coincidentally in the 1990s, as in the 1960s, many thought that technology could solve all problems—including social conflicts. The architectural forms that were invented in the 1990s—as those that were invented in the 1960s, or in the 1920s, for that matter—are a testament to that optimism. Today, that optimism is gone; and it shows. Digital designers today feel no attachment to the machines they use. Nobody likes robots. Not many today—regardless of age or generation—like the world as it is, either, or feel confident about its immediate prospects. The fragmentariness that is often the visual language of choice for many designers of today's digital avant-garde, and their occasional predilection for forms deliberately meant to "offend the eye,"[27] are a deliberate sign of dissent; regardless of their mode or ease of production, a protestation against the distress of our times.

Epilogue:
Being Post-Digital

Nobody knows what post-digital means. In an often cited article, first published in 2013, media scholar Florian Cramer reviewed a number of different meanings of the term, mostly—but not exclusively—related to a feeling of digital fatigue, or to the straightforward rejection of digital tools then widespread among a new generation of "post-digital" artists and creatives; Cramer's emblem of post-digitality was the meme of a hipster who sat in full sunlight on a park bench ostentatiously typing on a clunky mechanical typewriter.[1] The architect and educator Sam Jacob recently went one step further, advocating the "post-digital" use of digital tools to simulate the style of pre-digital ones. In particular, and famously, Jacob spearheaded the use of mainstream image editing software (such as Photoshop) to create a new genre of digital images designed to look like mechanical cut-and-paste compositions, which Jacob called "super-collages," and which he promoted as an antidote to the sleek smoothness of early computer renderings.[2] This strategy appealed to many architectural designers and theoreticians of different denominations, united as much by their ideological aversion to digital technologies as by their nostalgic affection for late mechanical ones: as discussed at length in chapter 4, collages, cut-and-paste, and visual (or verbal) citations are legacy compositional devices of the industrial-mechanical age that ended up being adopted by postmodernists of the late 1970s and 1980s. As a collage of cut-and-pasted chunks is bound to look chunky, chunkiness—both of the gaudy, ludic kind and in the sterner classicist mode, harkening back to Colin Rowe and historicist postmodernism—remains to this day the stylistic signifier of many designers of this ilk, and one of the distinctive visual trends of our time (early 2022).

However, as Cramer already pointed out back in 2013, "post-" is a polysemic prefix, which in the history of politics, philosophy, and the arts, may mean rejection as well as evolution and mutation. And indeed, unrelated to the regressive version of post-digitality just mentioned, some digital designers have themselves long been invoking a post-digital turn—but aiming at critical renewal of digitally intelligent design, not at a repudiation of it. After the phenomenal rise of computer-aided design and fabrication in the late 1990s, some of the very protagonists and leading theoreticians of the first digital turn in architecture didn't hesitate to give vent to their growing dissatisfaction, or even disillusionment: see for example the post-digital careers, so to speak, of Bernard Cache, or Lars Spuybroek. And, particularly after the financial crisis of 2007–2008, a new generation of digital designers (often trained by first generation blobmeisters) felt increasingly out of kilter with the technophilic, libertarian spirit of the digital 1990s. On aesthetic as well as purely technical grounds, many started to resent the reduction of digitally intelligent design to simplified spline-modeling and digital streamlining, and smooth curvilinearity, revolutionary when first heralded as an incarnation of the Deleuzian Fold, soon came to be seen as a sign of the excesses and increasingly unpalatable political climate of the 1990s. The political allegiances of some digital theorists soon started to appear suspicious, too: neoliberalism was popular in the immediate aftermath of the fall of the Berlin Wall, but right-wing libertarianism, the political ideology of choice for many venture capitalists that invested in, promoted, and profited from, digital technologies was way less favorably viewed after the financial implosion of 2007–2008, which digital technologies were in many ways instrumental in abetting.

Since then, these political preoccupations have been more than corroborated by the role of digital media in the rise of populism, and by the endorsement of populist or libertarian politics by some Silicon Valley magnates. While many experimental designers strived to distance themselves from the ideological and stylistic legacy of the first digital turn, in the course of the 2010s the unprecedented performance of some new technical tools led to a significant change of tack in digital design practices: the "big data" revolution and the revival of some core features of Artificial Intelligence—particularly machine learning—have shown that computers, alongside their proven but ancillary capacities as drawing machines, can sometimes help to solve otherwise intractable design problems. In retrospect, we can now also appreciate to what extent this "second" digital turn was still deeply permeated, inspired, and driven by the ideology of complexism: that's one of the arguments of this book. The digital turn in architecture was that rare accident of history: an epoch-making technical revolution driven by a powerful anti-scientific ideology.

Let me clarify: complexism is not against science: it is against modern science. Complexity theory is a postmodern science and, on its best days, a new kind of science. But it is a post-human science, which suits computers better than humans. The unfolding of the pandemic (at the time of this writing, early January 2022, still ongoing) has already proven—in case further proof by numbers was needed—that the ideology of complexism, when applied to human societies, is a recipe for catastrophe; I would also venture, without verifiable proof, that the same applies when we let complexism drive design—never mind the amount of electronic computation we may marshal for that purpose. With a few exceptions, we design for humans, even when we use post-human tools. This is where the need for a new, truly post-digital agenda appears most clearly. Our conceptual approach to electronic computation needs a new start—when we use computers for design purposes, for sure; and perhaps in general.

There is nothing wrong with modern science: the science we studied at school still allows us to describe, formalize, predict, and act upon most events that affect our daily lives. But today we also know what August Comte, for example, could neither have known nor admitted—namely that the method of modern science is not universal, and many natural, social, and technical phenomena lie outside of its ambit. Artificial Intelligence is one of them. Artificial Intelligence, and human intelligence, work in different ways. Our own scientific method neither applies to, nor describes, the mostly heuristic processes whereby computers solve problems—including problems we could not solve in any other way, which is why electronic computation so often outsmarts us. As it happens, complexity theory is very apt at describing some core processes of Artificial Intelligence, and very helpful at designing them. Good. But as human intelligence and Artificial Intelligence evidently inhabit quite different realms, let's keep it that way. Each to its trade: computer scientists now mostly agree that training computers to imitate the logic of human thinking is a wasteful and possibly doomed strategy. But having humans trying to imitate the way computers solve problems is worse than that—it's sheer madness.

On the other hand, if we let computers work in their own way, following their inner technical logic, the merger of post-scientific computation and automated manufacturing may at long last deliver a new way of making things—and with it, a new *modern* blueprint for a post-industrial world. We have long claimed that digital mass customization is cheaper, faster, smarter, and more environmentally sustainable than the industrial mass production of standardized wares; and that the electronic transmission of information is cheaper, faster, smarter, and more environmentally sustainable than the mechanical transportation of people and goods. The pandemic has proven not so much the unsustainability of the old industrial model—that was proven long ago—as the viability of the new post-industrial model: when we had to shut down centralized factories, airports, and offices, we did. We could do so because digital technologies now offer an alternative.

Once again, there is an anecdote I often tell my students to explain how the next wave of computational automation can lead to a new way of living, working, and making stuff. Imagine that you are a tomato farmer, and you want to choose the best technology to grow and sell your tomatoes at the lowest cost. The traditional, modernist solution would read, ideally, as follows: first, you should genetically modify your tomatoes, so they would grow to standard, easily manipulable shapes and ripen at the same pace. Let's assume that the best tomato shape for picking, sorting, packaging, and shipping is cubic. An industrial robotic arm, driven by a scripted algorithm, would then shake rows of identical cubic tomatoes, all in the same shape and at the same point of ripeness, cull them from their hydroponic vines, and push them onto a conveyor belt. Another industrial robotic arm would then pick these cubic tomatoes for boxing, labeling, and shipping. A truck would convey these wares to a logistics hub, from where they would be dispatched to fulfillment centers around the world by road, air, or rail.

The alternative: use a more versatile, adaptive technology to make the entire process fit the natural variations of a natural product. Let tomatoes be tomatoes. Let them grow as they always did. Then train an automated harvester, driven by robotic vision, sensors, and AI, to scan each tomato as my great-grandmother, I presume, would have done: picking each at the right point in time for putting in a basket and selling at the local market. My great-grandparents lived a few miles from town, so they commuted to the market square by carts and bicycle. Today, a fleet of electric self-driving vehicles could easily do the job; the local market, for what I know, is still there—if reduced to a caricature of what it used to be. Using computational automation, we could bring it back to its original function—providing fresh local food to local residents. 159

The next frontier of automation will not replace industrial workers repeating identical motions to execute the alienated notational work needed to produce standardized, one-size-fits-all, wasteful items of globalized mass production. That's already done and we shall need less and less of that. The next frontier of automation will beget a new kind of artisan workers carrying out unscripted, endlessly variable, inventive, and creative tasks to produce no more no less than the right amount of non-standard stuff we need: where we need it, when we need it, as we need it; made to specs, made on site, and made on demand. The microfactories of our imminent future will be automated artisan shops. And yes, just as plenty of wares we make today cannot be made that way, and will likely stop being made, some of today's workers will lose their jobs, because their jobs will disappear or will be taken over by machines. That has already happened many times since the start of the industrial revolution, and we now know that there are two ways to manage the social costs of innovation: some choose to protect those immediately hit and mete out the pains and gains of technical change across all sections of society as fairly as possible. Welfare states did so in different times, and today universal basic income, or other similar programs, might do that again. Countries that favor evolution by conflict and social Darwinism will likely succumb to the toll of civil unrest. But one thing is certain: while the adoption, or rejection, of some new socio-technical models will ultimately be a political choice, the merger of computation and post-industrial automation is no longer a vision for our future: as the climate crisis and the pandemic have shown, this may as well be the only future we have. And there may be a faint glimmer of hope in that designers, including some mentioned in this book, are already imagining that future—trying to figure it out, and, if we are lucky, finding ways to make it work.

5.1 Sabin Lab at Cornell University in collaboration with the DEfECT lab at Arizona State University, Agrivoltaic Pavilion, rendering, 2020. Customized, 3D printed Building Integrated Photovoltaics (BIPV) computationally designed for site-specific, non-mechanical tracking solar collection systems, inspired by studies of heliotropic mechanisms in sunflowers and light-scattering structures in Lithops plants. An integrated, sustainable approach to farming, energy generation and water conservation, offering an alternative to current solar technologies. Principal Investigators: Jenny Sabin & Mariana Bertoni. Research Team: Allison Bernett, Jeremy Bilotti, Begum Birol, Omar Dairi, Alexander Htet Kyaw, April Jeffries, Ian Pica Limbaseanu, Jack Otto. Image courtesy Sabin Lab, Cornell University.

Acknowledgments

The author gratefully acknowledges the support of the Architecture Research Fund of The Bartlett School of Architecture, University College London.

Notes

Chapter 1

1   Walter Benjamin, "The Storyteller: Reflections on the Works of Nikolai Leskov," in *Illuminations: Essays and Reflections*, trans. Harry Zohn, ed. Hannah Arendt (New York: Schocken Books, 1968), 83–109; first published as "Der Erzähler: Betrachtungen zum Werk Nikolai Lesskows," *Orient und Okzident*, n.s., 3 (October 1936): 16–33.

2   The primacy of recording over communication in the technical logic of the digital world is central to Maurizio Ferraris's theory of systemic hysteresis (in the philosophical, not in the technical sense of the term). See his recent *Documanità. Filosofia del mondo nuovo* (Rome; Bari: Laterza, 2021).

3   Some of these general arguments are variously mentioned in my earlier work: see in particular "Pattern Recognition," in *Metamorph. Catalogue of the 9th International Biennale d'Architettura, Venice 2004*, vol. 3, *Focus*, ed. Kurt W. Forster (Venice: Marsilio; New York: Rizzoli International, 2004), 44–58; and "Variable, Identical, Differential," in *The Alphabet and the Algorithm* (Cambridge, MA: MIT Press, 2011), 1–49.

4   See, for example, Alfred Sloan, *My Years with General Motors* (Garden City, NY: Doubleday, 1964), 20–21.

5   The locus classicus on the phonetic alphabet as a technology (today we would say a "cultural technology") is still Walter J. Ong, *Orality and Literacy: The Technologizing of the Word* (London: Methuen, 1982).

6   John Ruskin saw print as "the root of all mischief" (in this context, the original sin and the germ of all industrial evils to follow) because it is "that abominable art of printing" that "makes people used to have everything the same shape." Letter to Pauline Trevelyan, September 1854, in *The Works of John Ruskin*, ed. E. T. Cook and A. Wedderburn, vol. 36 (London: George Allen, 1909), 175.

7   By collating data reported by Febvre and Martin (1958), Eisenstein (2005), and Mack (2005), Alan Stangster recently concluded that in 1483 in Florence the unit cost of a printed copy of Marsilio Ficino's Latin translation of Plato's *Works* (on a print run of around 1,000 copies) would have been comprised between 1/300 and 1/400 of the cost of a scribal copy of the same text, and might have been sold at a price between ten to fifty times cheaper. Alan Sangster, "The Printing of Pacioli's Summa in 1494: How Many Copies Were Printed?" *The Accounting Historians Journal* 34, no. 1 (June 2007): 125–145, esp. 130–132. The classic study by Lucien Febvre and Henri-Jean Martin, *L'apparition du livre* (Paris: Albin Michel, 1958; see in particular chapter 7), offered the first assessment of the economics of mass production in early modern printing. A detailed study of the production costs of a 1486 incunable in Venice is in Michael Pollak, "Production Costs in Fifteenth-Century Printing," *Library Quarterly* 39, no. 4 (1969): 318–330.

8   James Womack, Daniel Jones, and Daniel Ross, *The Machine That Changed the World* (New York: Macmillan/Rawson Associates, 1990), 21–29. The Model T was produced from 1908 to 1927. In his autobiography, published

in 1922, Henry Ford recounts that "in 1909 I announced one morning, without any previous warning, that in the future we were going to build only one model, that the model was going to be 'Model T,' and that the chassis would be exactly the same for all cars, and I remarked: 'any customer can have a car painted any colour that he wants so long as it is black.'" Henry Ford, with Samuel Crowther, *My Life and Work* (Garden City, NY: Doubleday, Page & Co., 1922), 72. In 1914, after Ford famously raised his company's minimum wages to five dollars per day, a Model T cost less than four months' salary for his least-paid worker, and from 1910 to 1925 the price of the Model T declined from $900 to $260: https://corporate.ford.com/articles/history/moving-assembly-line.html.

9   See in particular "Maisons en série," in Le Corbusier-Saugnier, *Vers une architecture* (Paris: G. Crès et Cie, 1923), 185–224; first published as Le Corbusier-Saugnier, "Esthétique de l'ingénieur. Maisons en série," *L'Esprit nouveau* 13 (1921): 1525–1542; Le Corbusier, "Une ville contemporaine," in *Urbanisme* (Paris: G. Crès et Cie, 1925), 166–168.

10 Le Corbusier-Saugnier, *Vers une architecture*, 200–201; Le Corbusier-Saugnier, "Esthétique de l'ingénieur. Maisons en série," 1525–1542, 1538.

11 The *Exposition Internationale des Arts Décoratifs et Industriels Modernes* was held in Paris from April to October 1925.

12 Le Corbusier, *Urbanisme* (Paris: G. Crès et Cie, 1925), 263–274. On Corb's unsuccessful plea to André Citroën: ibid., 263–264, n. 1. To corroborate his theories Le Corbusier crammed his articles and his essays on urbanism of the early twenties with a wealth of dubious statistical data, meant to prove the irresistible rise of automobility in the US: see Carpo, "Post-Hype Digital Architecture. From Irrational Exuberance to

Irrational Despondency," *Grey Room* 14 (2004): 102–115.

13 On Le Corbusier's reading of Taylor, and on the influence of scientific management on modernist architectural theory, see an economist's take in Mauro F. Guillén, *The Taylorized Beauty of the Mechanical: Scientific Management and the Rise of Modernist Architecture* (Princeton, NJ: Princeton University Press, 2006).

14 Mario Carpo, *The Alphabet and the Algorithm*, 93–106. For an economist's view, see Chris Anderson, *Makers: The New Industrial Revolution* (New York: Random House Business Books, 2012), 81–89.

15 The idea of mass customization arose in the context of the postmodern pursuit of increased customer choice; a noted experiment in early mass customization was the multiple-choice model, based on a finite number of conspicuous but often purely cosmetic product variations. See Stanley M. Davis, *Future Perfect* (Reading, MA: Addison-Wesley, 1987), where the expression "mass customization" may first have occurred; Joseph B. Pine, *Mass Customization: The New Frontier in Business Competition*, with a foreword by Stan Davis (Boston: Harvard Business School Press, 1993). At some point in the course of the 1990s it became evident that digital design and production tools (then called CAD/CAM, or "file to factory" technologies) would allow for the mass production of endless variations, theoretically at no extra cost; in a sense, digital mass customization thus provided a technological answer to a longstanding postmodern quest for product variations. On the rise of the notion of "digital mass customization" in the 1990s see Mario Carpo, *The Second Digital Turn: Design Beyond Intelligence* (Cambridge, MA: MIT Press, 2017), 179–183.

16 Barring a sudden return to the Stone Age, the expectation is that the global costs of human labor will keep growing, and the global costs of computation and of robotic fabrication will keep declining—but that's a matter for another discussion, as it involves a number of social, political, and ethical issues. By one metric of the Organisation for Economic Co-operation and Development (OECD), the Unit Labor Cost per person employed (defined as "the average cost of labor per unit of output produced"), the costs of labor have risen in most OECD countries at an average of more than 4 percent per year between 2010 and 2020 (with some notable exceptions at both ends of the scale): see OECD (2021), "Unit labor costs" (indicator): https://doi.org/10.1787/37d9d925-en. Data from the International Labour Organization (ILO Department of Statistics, United Nations) on the "Mean nominal hourly labour cost per employee per economic activity" for the period 2009 to 2018 and for countries with high rates of robotic labor provide similar figures, indicating an annualized rise of human labor costs comprised between 3 and 5 percent (https://ilostat.ilo.org/topics/labour-costs/). A 2011 study by the Boston Consulting Group projected wages in China to rise 18 percent annually, reaching 6.31 USD in 2015 from a starting hourly wage of 0.72 USD in 2000: see Harold L. Sirkin, Michael Zinser, and Douglas Hohner, "Made in America Again: Why Manufacturing will Return to the US," Boston Consulting Group, 2011, https://image-src.bcg.com/Images/made_in_america_again_tcm9-111853.pdf. The International Labour Comparisons program reports that hourly compensation costs in manufacturing in China rose from 0.60 USD in 2002 to 4.11 USD in 2013, which would give an annualized rate of growth of more than 19 percent: "International Comparisons of Hourly Compensation Costs in Manufacturing, 2016—Summary Tables," The Conference Board, https://www.conference-board.org/ilcprogram/index.cfm?id=38269#Table2.

Robotic labor cost is notoriously difficult to appraise; for mainstream industrial applications it is currently (2021) estimated at less than five dollars per hour. Industrial robotic labor became cheaper than human labor in Germany in the 1980s, in the US in the 1990s, and in China around 2010 (Brian Carlisle, "Pick and Place for Profit: Using Robot Labor to Save Money," *Robotics Business Review*, September 22, 2017, https://www.roboticsbusinessreview.com/manufacturing/pick-place-profit-using-robot-labor-save-money). A 2015 report by the Boston Consulting Group compared the cost of labor of a human welder (estimated at $25 per hour including benefits) with the hourly cost of a robot carrying out the same tasks (estimated at $8 "once installation, maintenance, and the operation costs of all hardware, software, and peripherals are amortized over a five-year depreciation period"), and predicted that within fifteen years the operating cost per hour for a robot doing similar welding tasks "could plunge to as little as $2 when improvements in its performance are factored in." See: Michael Zinser, Justin Rose, and Hal Sirkin, "How Robots Will Redefine Competitiveness," Boston Consulting Group, September 23, 2015, https://www.bcg.com/publications/2015/lean-manufacturing-innovation-robots-redefine-competitiveness.

In recent times economists have also started assessing the costs of Robotic Process Automation (RPA),   167 i.e., the costs of the automation of white collar, or office work ("robotic" in this case does not refer to manufacturing tools but to pure data manipulation). RPA reduces the time humans spend dealing with information systems and doing repetitive tasks like typing and extracting data, formatting, copying, or generally moving information from one system to another. A rule of thumb calculation is that since RPA robots can work 24/7 they are estimated to replace the work of 1.7 humans, and to cut data entry costs to 70 percent: see Filipa Santos, Rúben Pereira, and José Braga Vasconcelos, "Toward Robotic Process Automation Implementation: An End-to-End Perspective," *Business Process Management Journal* 26, no. 2 (2020): 405–420, http://dx.doi.org/10.1108/BPMJ-12-2018-0380.

17 "Differential" reproduction, and non-standard serial production, were conceptualized in the early and mid-1990s, by Greg Lynn and Bernard Cache, respectively. See Carpo, *The Alphabet and the Algorithm*, 90–91, 130, n. 10.

18 See Chris Anderson, *Makers* (2012); Neil Gershenfeld, "How to Make Almost Anything: The Digital Fabrication Revolution," *Foreign Affairs* 91, no. 6 (November/December 2012): 43–57.

19 Cf. Carpo, "Republics of Makers," in *Imminent Commons: Urban Questions for the Near Future, Seoul Biennale of Architecture and Urbanism 2017*, ed. Alejandro Zaera-Polo and Hyungmin Pai (Barcelona: Actar, 2017), 302–309 (also available online: http://www.e-flux.com/architecture/positions/175265/republics-of-makers/); "Micro-managing Messiness: Pricing, and the Costs of a Digital Non-Standard Society," *Perspecta* 47, "Money" (2014): 219–226; Carpo, *The Second Digital Turn*, 153–157.

20 The Fablab (or Fab Lab) movement is to this day steered by the MIT Center for Bits and Atoms, which Gershenfeld founded in 2001. See Neil Gershenfeld, *Fab: The Coming Revolution on Your Desktop, from Personal Computing to Personal Fabrication* (New York: Basic Books, 2005).

21 Christina Cogdell has recently reviewed the existing scientific literature on the subject, based on a full life-cycle assessment (LCA) of contemporary tools for electronic computing. This takes into account the nature and provenance of raw materials used, a calculation of cumulative embodied energies (including energy consumed in operational use), and the costs of recycling and waste disposal. These thermodynamical and environmental loads do not even consider the costs of the infrastructure needed to keep computers running and connected to one another (electrical grid and communication networks), yet they suffice to demonstrate, in Cogdell's view, that digital technologies are "environmentally devastating" and "finite" (i.e., unsustainable). See Cogdell, *Toward a Living Architecture? Complexism and Biology in Generative Design* (Minneapolis, MN: University of Minnesota Press, 2018), 95–101. A 2019 report published by The Shift Project, a Paris-based non-profit think-tank, has shown that the share of digital technologies in global greenhouse gas emissions increased from 2.5 percent in 2013 to 3.7 percent in 2019 (which puts them on par with the aviation industry's emissions). The same report also estimated that the digital industry's energy use is increasing globally by around 4 percent a year, and in order to mitigate the resulting environmental costs called for a more parsimonious use of digital information and communication technologies ("digital sobriety"): see

Hugues Ferreboeuf et al., *Lean ICT: Towards Digital Sobriety* (March 2019), https://theshiftproject.org/wp-content/uploads/2019/03/Lean-ICT-Report_The-Shift-Project_2019.pdf. For similar reasons some media scholars now advocate curtailing energy-intensive media streaming in favor of a new "small-file aesthetics": see Laura U. Marks, "Let's Deal with the Carbon Footprint of Streaming Media," *Afterimage* 47, no. 2 (June 2020): 46–52, https://doi.org/10.1525/aft.2020.472009. A concurrent assessment of the energy consumption and greenhouse gas emissions due to the infrastructure of the internet is in Anne Cecile Orgérie and Laurent Lefèvre, "Le vrai coût energétique du numérique," *Pour la Science* 518 (December 2020): 48–59 (with bibliography).

For a more optimistic view of the energy consumption of data centers see: Masanet et al., "Recalibrating global data center energy-use estimates," *Science* 367, no. 6481 (February 28, 2020), 984–986, https://doi.org/10.1126/science.aba3758; on the energy consumption of Google searches, see James Glanz, "Google Details, and Defends, its use of Electricity," *The New York Times*, September 8, 2011; on the energy consumption and e-waste (discarded electronic equipment) generated by bitcoin mining, see Alex de Vries and Christian Stoll, "Bitcoin's Growing E-Waste problem," *Resources, Conservation & Recycling* 175 (2021), 105901, https://doi.org/10.1016/j.resconrec.2021.105901. For a more general assessment of e-waste see Elizabeth Grossman, *High Tech Trash: Digital Devices, Hidden Toxics, and Human Health* (Washington, DC: Island Press/Shearwater Books, 2006); Jennifer Gabrys, *Digital Rubbish: A Natural History of Electronics* (Ann Arbor, MI: University of Michigan Press, 2011); Patrick Brodie and Julia Velkova, "Cloud ruins: Ericsson's Vaudreuil-Dorion data centre and infrastructural abandonment," *Information, Communication & Society* 24, no. 6 (April 2021): 869–885.

22 On mass production versus customization, see below in this chapter; the advantages of the electronic transmission of data versus the mechanical transportation of people and goods would appear self-evident: how would a phone call not be economically, environmentally, and energetically preferable to a sixteen-hour intercontinental flight? Yet many scholars and scientists argue that they can prove the opposite; see for example the frequent claim that "the internet produces more greenhouse gases than aviation" (verbatim in *Pour la Science*, 457, October 21, 2015, but the same or very similar claims are ubiquitous in scientific literature, in popular science, and in the general press). Likewise, prestigious academic research suggests that office workers use more electricity and emit more greenhouse gases when working remotely from their bedrooms than when commuting to work by train or car: see, for example, Andrew Hook et al., "A systematic review of the energy and climate impacts of teleworking," *Environmental Research Letters* 15 (2020): 093003 (https://doi.org/10.1088/1748-9326/ab8a84).

23 Carpo, *The Second Digital Turn*, 56–59.

24 The production of I-beams in a traditional, mid-twentieth-century steel mill was not strictly speaking a matrix-based process, unless the beam was extruded through a die; however, the same logic of scale applied to the more common hot-rolled and cold-rolled I-beams, because rolled beams are made through a continuous casting and milling operation that, just like an extrusion, produces the same profile throughout; resetting the rolling

169

mills would require stopping and restarting the process, hence entail additional costs, similar to replacing a mold in a matrix-based production process.

25 Jean-Pierre Cêtre, "Neue Nationalgalerie recto-verso," *Faces: Journal d'architecture*, "Tectonique (I)" 47 (1999–2000): 34–41.

26 See note 16.

27 See Jeremy Rifkin, *The Zero Marginal Cost Society: The Internet of Things, The Collaborative Commons, and the Eclipse of Capitalism* (New York: Palgrave Macmillan, 2014); and, on zero cost robotics, Nick Srnicek and Alex Williams, *Inventing the Future: Postcapitalism and a World Without Work* (London: Verso, 2015), in particular chapter 6, "Post Work Imaginaries"; Aaron Bastani, *Fully Automated Luxury Communism: A Manifesto* (London: Verso, 2019).

28 An intriguing counterargument has been made recently by Manja van de Worp, the engineer of some of Gilles Retsin's "neo-brutalist" work, who has claimed that a redundant, oversized, hyper-generic structure can be more environmentally friendly and, in the long term, more convenient than a carefully customized one, because alongside the specific loads for which it was designed, a heavier generic structure may withstand different, and unforeseeable, future conditions of use: Manja van de Worp, "Rubens Structures: A Different Lightness Through Performance Adaptability," in Gilles Retsin, ed., "Discrete: Reappraising the Digital in Architecture," special issue (*AD* Profile 258), *Architectural Design* 89, 2 (2019): 54–59. On a similar note, Gilles Retsin has argued that oversized, generic structures are in fact beneficial for the environment, if the material being wasted is timber, because timber is a "carbon sink," hence the more timber we put into building, regardless of purpose, the more carbon dioxide we sequester from the atmosphere (so long as said timber never rots or burns): Retsin, "Toward Discrete Architecture: Automation Takes Command," paper presented at *ACADIA 2019*, *Ubiquity and Autonomy*, University of Texas at Austin, School of Architecture, October 24–26, 2019, http://papers.cumincad.org/cgi-bin/works/Show?_id=acadia19_532. It may also be noted that oversized structures today are often required for fireproofing, unrelated to load-bearing performance.

29 Karl Marx's theory of alienation (estrangement, *Entfremdung* in Marx's original) is outlined in his *Economic and Philosophic Manuscripts of 1844*: see in particular the *First Manuscript*, chapters 23–24. Marx defines several modes of alienation induced by the economic separation between the worker and the ownership of the means of production, which characterizes capitalism, but also by the "physical and mental" separation between the worker and the technical logic of new modes of production, which characterizes the industrial system.

30 An earlier version of this section was published as "Coronavirus Might Give Us the Internet We Always Wanted," *The Architect's Newspaper*, April 13, 2020, https://www.archpaper.com/2020/04/coronavirus-might-give-internet-weve-always-wanted/; reposted as "Postcoronial Studies: Coronavirus, and the Return of the Internet," *Labont Blog*, Center for Ontology, Università di Torino, April 16, 2020, labontblog.com/2020/04/16/postcoronial-studies-coronavirus-and-the-return-of-the-internet/; and elsewhere with different titles. The expression "Post-coronial" is, to my knowledge, an invention of Professor Maurizio Ferraris (Center for Ontology of the University of Turin); see his best-selling *Post-Coronial*

Studies. Seicento Sfumature di Virus (Turin: Einaudi, 2021).

31 William J. Mitchell, *City of Bits: Space, Place, and the Infobahn* (Cambridge, MA: MIT Press, 1995).

32 Joseph Rykwert, *The Seduction of Place: The City in the Twenty-First Century* (New York: Pantheon Books, 2000), 156.

33 Mark C. Taylor and Esa Saarinen, *Imagologies: Media Philosophy* (London: Routledge, 1994).

34 Corinne Le Quéré et al., "Temporary reduction in daily global CO_2 emissions during the COVID-19 forced confinement," *Nature Climate Change* 10 (2020): 647–653, https://doi.org/10.1038/s41558-020-0797-x. This study was submitted on April 9, 2020, only a few weeks into the global lockdowns; its results were partly revised by later research.

35 For a palette of different design positions on climate change, from techno-optimist to collapsologist, see Elisa Iturbe, ed., "Overcoming Carbon Form," *Log* 47 (December 2019). The term "collapsology" comes from Pablo Servigne and Raphaël Stevens's best-selling *Comment tout peut s'effondrer: Petit manuel de collapsologie à l'usage des générations présentes*, first published 2015, and translated into English in 2020 as *How Everything Can Collapse: A Manual for Our Times*, trans. Andrew Brown (Medford, MA; Cambridge: Polity, 2020).

36 In a different context, similar arguments have been made by Bruno Latour in an article published March 30, 2020, on the French digital platform AOC, "Imaginer les gestes-barrières contre le retour à la production d'avant-crise": https://aoc.media/opinion/2020/03/29/imaginer-les-gestes-barrieres-contre-le-retour-a-la-production-davant-crise. For an English translation, see Latour's website: http://www.bruno-latour.fr/fr/node/849.

37 Jean-Louis Cohen, *Architecture in Uniform: Designing and Building for the Second World War* (Montreal: Canadian Centre for Architecture; Paris: Hazan; 2011).

Chapter 2

1 For an earlier discussion of some of the topics under review in this chapter see Mario Carpo, "Rise of the Machines: Mario Carpo on Robotic Construction," *Artforum* 58, no. 7 (2020): 172–179.

2 On the rise of a "second" digital style of Big Data and discreteness (subsequent to, and in many ways a reaction against, the "first" style of digital streamlining, or spline-modeling, in the 1990s) see Mario Carpo, "Breaking the Curve: Big Data and Digital Design," *Artforum* 52, no. 6 (2014): 168–173; and *The Second Digital Turn: Design Beyond Intelligence* (Cambridge, MA: MIT Press, 2017).

3 More precisely, a tea and coffee service. See Greg Lynn, "Variations calculées," in *Architectures non standard*, ed. Frédéric Migayrou and Zeynep Mennan (Paris: éditions du Centre Pompidou, 2003), 91. According to commercial information furnished by manufacturer Alessi SpA, the original project included fifty-thousand variations, of which ninety-nine were made in addition to the author's three copies.

4 Greg Lynn, "Embryologic Houses," in Ali Rahim, ed., "Contemporary Processes in Architecture," special issue (*AD* Profile 145), *Architectural Design* 70, no. 3 (May–June 2000): 26–35; Mario Carpo, *The Digital Turn in Architecture, 1992–2012* (Chichester: Wiley, 2013), 125–131; Howard Schubert, "Embryological House," http://www.cca.qc.ca/en/issues/4/origins-of-the-digital/5/embryological-house.

5 Neil Leach et al., "Robotic Construction by Contour Crafting: The Case of Lunar Construction," *International Journal of Architectural*

Computing 10, no. 3 (September 2012): 423–438.

6 Neil Leach, "Introduction," in *Digital Fabrication*, ed. Philip F. Yuan, Achim Menges, Neil Leach (Shanghai: Tongji University Press, 2017), 13–27: 15, with regard to the 3D printing of jewelry, *haute couture*, and wedding cakes (but Leach does not apply the same limitations to other modes of digital fabrication).

7 As an illustration of the open-ended, collaborative nature of major civic building projects in a medieval city-state, see my summary of the history of the Cathedral of Florence (and of Brunelleschi's disruption of the local, guild-based traditions) in *The Alphabet and the Algorithm*, 71–79 (with bibliography).

8 In the dedication to Brunelleschi ("Pippo architetto") of the Italian version of his book *On painting* (1435–36): Leon Battista Alberti, *De Pictura*, in Alberti, *Opere volgari*, vol. 3, ed. Cecil Grayson (Laterza: Bari, 1973), 7–8; Leon Battista Alberti, *On Painting and On Sculpture: The Latin Texts of* De Pictura *and* De Statua, ed. and trans. Cecil Grayson (London: Phaidon, 1972), 32–33.

9 During the eighteen years of his service as the head of works for the construction of the dome Brunelleschi's official title was *provveditore* and occasionally *governatore* (both terms roughly translate as "supervisor"). The term "inventor," used by Brunelleschi's biographer Antonio Manetti and then by Vasari, is not attested in the books of the Opera del Duomo: see Margaret Haines, "Myth and Management in the Construction of Brunelleschi's Cupola," *I Tatti Studies in the Italian Renaissance* 14–15 (2011–12): 47–101. When, a few months before the dedication of the dome, the Officers of the Works once tried to call Brunelleschi an "architect," they grossly misspelled the word, which evidently they were not familiar with (Haines, 55). To be noted, however, Giannozzo Manetti's chronicle of the 1436 consecration of the Cathedral by Pope Eugene IV never mentions Brunelleschi: see Caroline Van Eck, "Giannozzo Manetti on Architecture: *The Oratio de secularibus et pontificalibus pompis in consecratione basilicae Florentinae of 1436*," *Renaissance Studies* 12, no. 4 (1998): 449–475.

10 Brunelleschi did, however, obtain a three-year patent—issued on June 19, 1421, and likely the first modern patent in history—for an unusually big ship he had devised to carry heavy loads over the Arno, and which he meant to exploit to sell marble for the building of the dome to the Office of the Works which employed him. Like the *Titanic*, Brunelleschi's monster ship (as it was nicknamed at the time) sunk on her maiden voyage. Ludwig H. Heydenreich, *Architecture in Italy 1400–1500* (New Haven, CT; London: Yale University Press, 1996), 13, n. 7; first published Harmondsworth: Penguin Books, 1974; Frank D. Prager, "Brunelleschi's Inventions and the Revival of Roman Masonry Work," *Osiris* 9 (1950): 457–507; Frank D. Prager, "Brunelleschi's Patent," *Journal of the Patent Office Society* 28, no. 2 (February 1946): 109–35.

11 On the rise of the "Albertian paradigm" (architecture as an authorial, allographic, notational art, built by design and designed by notation) see Carpo, *The Alphabet and the Algorithm*, 71–77; "Craftsman to Draftsman: The Albertian Paradigm and the Modern Invention of Construction Drawings," in *The Working Drawing*, ed. Annette Spiro and David Ganzoni (Zurich: Park Books, 2013), 278–281. Similar conclusions, from a different standpoint, in Marvin Trachtenberg, *Building in Time: From Giotto to Alberti and Modern Oblivion* (New Haven, CT; London: Yale University Press, 2010).

12 Leon Battista Alberti, *De re aedificatoria*, 9.11.1: "Dignitatem idcirco servasse consulti est; fidum consilium poscenti castigataque lineamenta praestitisse sat est." ("A wise man will stand on his dignity; save your sound advice and fine drawings for someone who really wants them.") Alberti, a gentleman designer, refers to non-remunerated design work ("nulla repensa gratia"), but elsewhere in the same paragraph he seems to imply that design work should receive pay. Latin text from Leon Battista Alberti, *L'architettura [De re aedificatoria]*, ed. and trans. Giovanni Orlandi (Milan: Il Polifilo, 1966), 863; references by book, chapter, and paragraph are to Orlandi's Latin edition. English translation from Leon Battista Alberti, *On the Art of Building in Ten Books*, trans. Joseph Rykwert, Neil Leach, and Robert Tavernor (Cambridge, MA: MIT Press, 1988), 318. Witness the patents he obtained, and his struggles with the Cathedral commissioners on pay, Brunelleschi was more business-minded (but he was paid a salary, not fees, by the Office of the Works).

13 In the same order: Alberti, *De re aedificatoria*, 2.4.1; 9.10.3; 9.9.6; 9.11.2; English trans. (1988), 38, 315, 314, 318, n. 125.

14 The Ciompi revolt (*tumulto dei Ciompi*), 1378–1382.

15 See Alberti's famous letter to Matteo de' Pasti, dated November 18, 1454, sent from Rome to Rimini. Alberti had provided a design for the church of San Francesco in Rimini, which was being built in his absence, and he vociferously complained that his original "model" and "drawing" were being disregarded by the local builders. For an English translation of the letter, see Robert Tavernor, *On Alberti and the Art of Building* (New Haven, CT; London: Yale University Press, 1998), 60–63, n. 64.

16 Alberti does not seem to have ever doubted that building materials would behave in some predictable fashion, but he had no way to prove that with the science of his time. Book II of *De re aedificatoria* was entirely devoted to materials, but the book is a bland doxographic compilation from classical sources, mostly Pliny, Vitruvius, Cato, and Varro, with only a couple of personal remarks added. The book opens with one of the strongest invocations of the Albertian notational mandate—with Alberti reiterating the need to abstain from any design change after construction starts, and detailing the geometrical specifications of the notational drawings needed to make his predictive design system work. The book concludes with an unrequested tirade against the "frivolous follies" of "idle superstitions," likely targeting the pagan rituals of divination and auguration. Evidently, Alberti had no sympathies for heathen pantheism. The foundations of a new science for the prediction of the mechanical behavior of inert materials would have to wait a bit longer; Galileo, who inaugurated it in his last book, *Discorsi e Dimostrazioni Matematiche* (1638) started his teaching career in the Florentine Academy of the Arts of Drawing, Vasari's brainchild, which institutionalized Alberti's humanistic invention of the three "fine" arts.

17 Frederick Winslow Taylor, *The Principles of Scientific Management* (New York: Harper, 1911), 24–25, 31. All citations to follow from the first edition.

18 Ibid., 37, 82–85.

19 Ibid., 59.

20 Ibid., 26, 41.

21 Ibid., 41.

22 Ibid., 59–62.

23 Ibid., 14, 21, 50.

24 Ibid., 27.

25 Ibid., 39.

26 Ibid., 59.

27 Ibid., 119.

28 Ibid., 126.

29 Ibid., 77–81, 113, 117, etc. Taylor refers to Gilbreth's book on bricklaying, which had been published in 1909.

30 M. J. Steel, D. W. Cheetham, "Frank Bunker Gilbreth: Building Contractor, Inventor and Pioneer Industrial Engineer," *Construction History* 9 (1993): 51–69.

31 Frank B. Gilbreth, *Bricklaying System* (New York: Myron C. Clark; London: Spon, 1909), 140.

32 Ibid., 148–153. Gilbreth's motion studies preceded Rudolf von Laban's, whose notational system for choreography (*Schrifttanz*, or Written Dance) was published in 1928; later in life Laban wrote in turn a handbook of time and motion studies, published in 1947. The term "ergonomics," which today would cover similar matters, was only coined in 1949.

33 Frank B. Gilbreth, *Motion Study. A Method for Increasing the Efficiency of the Workman* (New York: Van Nostrand, 1911), 67.

34 Ibid., 73.

35 This story, and others, are told in the best-selling memoir *Cheaper by the Dozen* (1948), written by Frank B. Gilbreth Jr and Ernestine Gilbreth Carey, two of the Gilbreths' twelve kids. A 1950 film with the same title, starring Clifton Webb and Myrna Loy, was only loosely based on the original story.

36 Gilbreth, *Bricklaying System*, 100.

37 Gilbreth, *Motion Study*, 129; *Bricklaying System*, 72, 91.

38 Gilbreth, *Bricklaying System*, 8–10.

39 Michael Peterson, "Thomas Edison's Concrete Houses," *Invention and Technology* 11 (Winter 1996): 54–55; Matt Burgermaster, "Edison's 'Single-Pour System': Inventing Seamless Architecture," paper presented at the 64th Annual Meeting of the Society of Architectural Historians, New Orleans, April 15, 2011; Randall Stross, *The Wizard of Menlo Park: How Thomas Alva Edison Invented the Modern World* (New York: Random House, 2007), 241–242, 263–264. Edison patented his poured concrete system in 1917.

40 Frederick W. Taylor and Sanford E. Thompson, *A Treatise on Concrete, Plain and Reinforced. Materials, Construction, and Design of Concrete and Reinforced Concrete* (Norwood, MA: Plimpton Press, 1905; repr. New York: Wiley; London: Chapman, 1907; 2nd ed. [expanded] New York: Wiley, 1909; repr. 1911, 1916 [expanded], 1925, 1931, 1939 [expanded]).

41 Taylor, *Concrete* (1905), 11–12. There is no cost analysis in any edition of Taylor's treatise, but Taylor and Thompson wrote a separate manual specifically on *Concrete Costs* (New York: Wiley, 1912); see Michael Osman, "The Managerial Aesthetics of Concrete," *Perspecta* 44, "Agency" (2012): 67–76.

42 Taylor, *Concrete* (1905), 282.

43 Taylor, *Concrete* (1909), 399.

44 Except in Italy, where it was equal to 10 from a Royal Decree of November 16, 1939 to a Ministerial Decree of January 9, 1996.

45 Taylor, *Concrete* (1909), 534.

46 Karl-Eugen Kurrer, *The History of the Theory of Structures: From Arch Analysis to Computational Mechanics* (Berlin: Ernst and Sohn, 2008), 496–525.

47 Taylor, *Concrete* (1909), 534.

48 Taylor, *Concrete* (1905), 21; (1909), 21.

49 Henry Ford, with Samuel Crowther, *My Life and Work* (Garden City, NY: Doubleday, 1922), 81.

50 Ibid., 80.

51 David A. Hounshell, "The Same Old Principles in the New Manufacturing," *Harvard Business Review* (November, 1988): https://hbr.org/1988/11/the-same-old-principles-in-the-new-manufacturing.

52 David A. Hounshell, "Ford Automates: Technology and Organization in Theory and Practice," *Business and*

Economic History 24, no. 1 (Fall 1995): 59–71; Margaret Wiley Marshall, "'Automation' Today and in 1662," *American Speech* 32, no. 2 (May 1957): 149–151.

53 Marshall, "'Automation' Today," 150.

54 Dimitrios Kalligeropoulos and Soultana Vasileiadou, "The Homeric Automata and Their Implementation," in *Science and Technology in Homeric Epics*, ed. S. A. Paipetis (New York: Springer, 2008): 77–84.

55 Aristotle, *Politics*, 1253b 20. Soultana Vasileiadou and Dimitrios Kalligeropoulos, "Myth, theory and technology of automatic control in ancient Greece," in *2007 European Control Conference (ECC)*, 249–256, https://doi.org/10.23919/ECC.2007 .7068430.

56 First explicitly formulated in Asimov's short story "Runaround," first published in *Astounding Science Fiction* (April 1942): 94–103.

57 Lisa Nocks, *The Robot: The Life Story of a Technology* (Baltimore, MD: John Hopkins University Press, 2008), 65–67; Philippe Morel, "Computation or Revolution," in Fabio Gramazio and Mathias Kohler, ed., "Made by Robots," special issue (*AD* Profile 229), *Architectural Design* 84, no. 3 (2014): 76–87, with a reproduction of Devol's original patent (United States Patent 2,988,237 Programmed Article Transfer, George C. Devol, Jr., Brookside Drive, Greenwich, Conn. Filed Dec. 10, 1954, Ser. No. 474,574 28 Claims. Cl. 214-11), granted 06.13.1961, https://patents .google.com/patent/US2988237 A/en; Joseph F. Engelberger, *Robotics in Practice: Management and Applications of Industrial Robots* (New York: Amacom, 1980), xv–xvii.

58 The first Unimate was delivered to a General Motors factory in New Jersey, where it was used to unload burning hot metal parts from a die casting machine: Engelberger, *Robots in Practice*, 13.

59 Ibid., 15.

60 Morel, "Computation or Revolution," 80–82.

61 Devol's original patent application (US 2,988,237) mentions jaws, a suction gripper, or "other comparable article handling tool."

62 The 2012 revisions of ISO entry 8373 add that robots should be able to "move within their environment"—a requirement then immediately waived for industrial robots, which can be "fixed in place"; to the contrary, service robots in particular (i.e., robots used outside of factories) appear exempt from the anthropomorphic mandate if they can offer some degree of automotivity. As a result, if we try to figure out the original vision behind the ISO nomenclatures, it would appear that robots today should either be stationary but then gesticulate like humans, or should be able to move around on their own, regardless of gesticulation; which is the same as saying that robots should have moving arms or moving legs (or be equipped with any prosthetic device that may replace the function of moving limbs). For a discussion of various definitions of robots see also Lisa Nocks, *The Robot*, 71–72.

63 George C. Devol Jr., US Patent 2,988,237.

64 Autocorrecting servomechanisms were the springboard of Wiener's new science of cybernetics (1948), originally seen as the study of the feedback loop that would correct an accidental variation in the direction of sailing set by a steersman, by automatically adjusting the angle of the rudder. See chapter 3, notes 23–25. Many analog or electromechanical amplifiers of motion, without any "cybernetic" feedback or correction, are often also called "servomechanisms."

65 John J. Craig, *Introduction to Robotics: Mechanics and Control*, 3rd rev. ed. (Upper Saddle River, NJ: Pearson

Prentice Hall, 2005), 12–14, chapter 10, 290–317 (first ed., Reading, MA: Addison Wesley, 1986); Craig, *Adaptive Control of Mechanical Manipulators* (Reading, MA: Addison-Wesley, 1988).

66 Maurice I. Zeldman, *What Every Engineer Should Know about Robots* (New York: Dekker, 1984), 38.

67 Wesley Stone, "The History of Robotics," in *Robotics and Automation Handbook*, ed. Thomas R. Kurfess (Boca Raton: CRC Press, 2005), 1–12; Nocks, *The Robot*, 100–105.

68 In fact, Engelberger argued that the best use of industrial robots would not come from making robots capable of responding to random incidents and unscripted workflow variations, but from eliminating all incidents and unscripted variations from the factory floor ("robots, to become ubiquitous, need rationalization of the factory"): Engelberger, *Robots in Practice*, 128.

69 Isaac Asimov, "Foreword," in Engelberger, *Robots in Practice*, xiii.

70 Jasia Reichardt, *Robots: Fact, Fiction and Prediction* (London: Thames & Hudson, 1978), 163.

71 Ibid., 56. Reichardt shows many art projects—from Harold Cohen's *Aaron* (1973–77), a rule-based, computer-driven drawing machine, to Edward Ihnatowicz's *Senster* (1970), a cybernetic robot that moves in reaction to people getting close to it—that would have kept the old dream alive, even at a time when industrial engineers and computer scientists alike seemed to have given up research in intelligent robotics.

72 An earlier version of parts of this section in "Rise of the Machines: Mario Carpo on Robotic Construction," *Art Forum* 58, no. 7 (2020): 172–179, 235.

73 The story told by Wesley Stone in Kurfess, *Robotic and Automation Handbook* (2005, 1–12), basically ends around 1980; the encyclopedic history of robotics and automation by Lisa Nocks (*The Robot*, 2008) did include further developments after the rise of industrial robotics in the 1960s and 1970s, but mostly outside of manufacturing, in fields like service robotics (vacuum cleaners, lawn mowers) and autonomous vehicles (95–160).

74 Fabio Gramazio, Matthias Kohler, and Jan Willmann, *The Robotic Touch: How Robots Change Architecture* (Zurich: Park Books, 2014), 15–22; Fabio Gramazio, Matthias Kohler, and Jan Willmann, "Authoring Robotic Processes," in Fabio Gramazio and Mathias Kohler, ed., "Made by Robots," special issue (*AD* Profile 229), *Architectural Design* 84, no. 3 (2014): 14–21.

75 That may be one reason why experiments in robotic automation meant to mass-produce simplified straight walls never caught on: see the ROCCO Robotic Bricklayer, developed in Germany circa 1994: Jurgen Andres, Thomas Bock, Friedrich Gebhart, Werner Steck, "First Results of the Development of the Masonry Robot System ROCCO," in *Proceedings of the 11th ISARC* (Brighton, UK, 1994), 87–93, http://doi.org/10.1016/B978-0-444-82044-0.50016-3; Gramazio, Kohler and Willmann, "Authoring Robotic Processes," 21); or similar projects by the Obayashi Corp in Japan, and the recent SAM Project in the US: Jonathan Waldman, *SAM: One Robot, a Dozen Engineers, and the Race to Revolutionize the Way We Build* (New York: Avid Reader Press, 2020). The economics of robotic bricklaying thus vindicates some basic tenets of digital design theory: if design simplicity and design complexity come at the same cost per brick laid, the use of bricklaying robots in the simple pursuit of economies of scale cannot compete against cheaper modes of standardized construction.

76 Gramazio et al., *The Robotic Touch*, 250–59.

77 See the Agile Robotics for Industrial
Automation Competition, ARIAC,
started 2017, organized by the
National Institute of Standards
and Technology (NIST) with the
support of the Open Source Robotics
Foundation, https://www
.nist.gov/el/intelligent
-systems-division-73500/agile
-robotics-industrial-automation
-competition.

78 Emmanuel Afrane Gyasi, Heikki
Handroos, Paul Kah, "Survey on
artificial intelligence (AI) applied in
welding," Conference on Flexible
Automation and Intelligent
Manufacturing (FAIM2019), June 24–
28, 2019, Limerick, Ireland, in *Procedia
Manufacturing* 38 (2019): 702–714, doi.
org/10.1016/j.promfg.2020.01.095.

79 I am grateful to Professor Enrico
Terrone of the University of Genoa
for some important clarifications on
the ontological difference between
notational and post-notational
automation.

80 Gramazio et al., *The Robotic Touch*,
84–99, 286–302, 422–429.

81 Carpo, *The Second Digital Turn*, 40–55.

82 Achim Menges, "The New Cyber-
Physical Making: Computational
Construction," in Menges, ed.,
"Material Synthesis. Fusing the
Physical and the Computational,"
special issue (*AD* Profile 237),
Architectural Design 85, no. 5 (2015):
28–33.

83 Achim Menges, "Coalescences of
machine and material computation,"
in *Digital Fabrication*, ed. Philip F.
Yuan, Achim Menges, Neil Leach
(Shanghai: Tongji University Press,
2017), 31–41: 37; "Future Wood,"
ibid., 115–122: 116.

84 Gramazio et al., "Rubble
Aggregation," in *The Robotic Touch*,
422–29; Martin Self and Emmanuel
Vercruysse, "Infinite Variations,
Radical Strategies. Woodchip Barn,"
Fabricate 2 (2017): 30–35; Achim
Menges, "Material Resourcefulness:
Activating Material Information in

Computational Design," in Menges,
ed., "Material Computation: Higher
Integration in Morphogenetic
Design," special issue (*AD* Profile
216), *Architectural Design* 82, no. 2
(2012): 42.

85 Kirstin Petersen and Radhika
Nagpal, "Complex Design by
Simple Robots: A Collective
Embodied Intelligence Approach
to Construction," in Skylar Tibbits,
ed., "Autonomous Assembly:
Designing a New Era of Collective
Construction," special issue (*AD*
Profile 248), *Architectural Design* 87, no.
4 (2017): 44–49; Michael Rubenstein
et al., "Kilobot: A Low Cost Robot
With Scalable Operations Designed
for Collective Behaviors," *Robotics
and Autonomous Systems* 62, no. 7 (July
2014): 966–975, https://doi
.org/10.1016/j.robot.2013.08.006.
However, the zoomorphic analogy
in this instance can be misleading:
while some insects can "self-
organize" to build their own abode
(termite mounds, anthills, etc.), a
colony of robots would likely be
tasked to build a house for someone
else, and would thence need to get
directions—or, indeed, design
instructions—from external sources.

86 Mollie Claypool, Manuel Jimenez
Garcia, Gilles Retsin, and Vicente
Soler, *Robotic Building: Architecture
in the Age of Automation* (Munich:
Detail, 2019), 74–83; Retsin,
"Toward Discrete Architecture:
Automation Takes Command,"
in *Acadia 2019: Proceedings of the 39th
Annual Conference*, University of Texas
at Austin, October 2019, 532–541,
papers.cumincad.org/cgi-bin/works
/Show?_id=acadia19_532.

87 Benjamin Jennett et al., "Material–
Robot System for Assembly of
Discrete Cellular Structures," *IEEE
Robotics and Automation Letters* 4,
no. 4 (October 2019): 4019–4026.
Contrary to the technical logic of
intelligent, post-notational robotics,
Gershenfeld's system, meant for

use in zero gravity and in outer space, is entirely deterministic: all motions are scripted and the physical operations carried out on site do not provide any feedback to the program (other than confirmation of execution).

88 Claypool et al., *Robotic Building* (2019); Retsin, ed., "Discrete: Reappraising the Digital in Architecture," special issue (*AD Profile 258*), *Architectural Design* 89, no. 2 (2019).

89 Mario Carpo, "Particlized," in Retsin, ed., *Architectural Design* 89, no. 2 (2019): 86–93. "Particlization" is Kengo Kuma's own term, first recorded in the English translations of some of his works from the early 2000s: see in particular: Kengo Kuma, *Anti-Object: The Dissolution and Disintegration of Architecture*, trans. Hiroshi Watanabe (London: The Architectural Association, 2008); first published in Japanese, 2000 (the original title in Japanese roughly translates as *Anti-Object: Melting and Crushing Buildings*); and Kengo Kuma, *Kengo Kuma: Materials, Structures, Details* (Basel: Birkhäuser, 2004); first published in Japanese, 2003.

Chapter 3

1 Clapeyron's theorem on elastic deformation work was published in 1858: E. Clapeyron, "Mémoire sur le travail des forces élastiques dans un corps solide élastique déformé par l'action de forces extérieures," *Comptes Rendus de l'Académie des Sciences*, 46 (1858): 208–212. Clapeyron was a professor at the École nationale des ponts et chaussées from 1844 to 1859, and he died in 1864.

2 To this day I could not ascertain if my friend's conclusions were right (even assuming that Clapeyron's theorem could be applied to snow, and furthermore making abstraction of all non-elastic aspects of the problem). Maurizio Brocato, professor of structural engineering at the École des Ponts ParisTech, and my colleague at the School of Architecture of Paris-Malaquais, thinks that in that dispute, as recounted here, I may have been less wrong than my friend. But as I believed my friend's explanation prima facie back then, that's irrelevant to the rest of the story I am telling here.

3 Adhémar Jean-Claude Barré de Saint-Venant's *Mémoire sur la torsion des prismes* is dated differently, from 1847 to 1856; Saint-Venant published several essays with similar titles, all including versions of his "principle": see Kurrer, *The History of the Theory of Structures* (2008), 399.

4 Gutenberg–Richter Law: see Per Bak, *How Nature Works: The Science of Self-Organized Criticality* (New York: Springer, 1996), 12–14, and passim.

5 Melanie Mitchell, *Complexity: A Guided Tour* (New York; Oxford: Oxford University Press, 2009), 240–245, 261, and passim.

6 George Kingsley Zipf, *Selected Studies of the Principle of Relative Frequency in Language* (Cambridge, MA: Harvard University Press, 1932); Mitchell, *Complexity*, 269–272.

7 Pareto's principle, or Pareto's law of unequal distribution, was first outlined in Vilfredo Pareto, *Cours d'économie politique professé à l'Université de Lausanne* (Lausanne: Rouge; and Paris: Pichon, 1896–97). Among the various instantiations of his law, Pareto had shown with facts and figures that 80 percent of the wealth of some nations was then in the hands of 20 percent of their respective populations; that 80/20 distribution has since inspired many "pop" versions of Pareto's law, stating for example that 80 percent of all sales come from 20 percent of clients, 80 percent of cases of marital infidelity come from 20 percent of husbands or wives, 80 percent of internet traffic is generated by 20 percent of websites, 80 percent of car

accidents come from 20 percent of drivers, and, in general, 80 percent of consequences come from 20 percent of causes.

8   Bak, *How Nature Works*, 14–24.

9   That is the main thesis of his seminal *How Nature Works* (partly published as articles in 1987–88). Bak thought that a universal regularity underpins the power laws of many natural and social phenomena alike, and called this single underlying rule "self-organized criticality": "large avalanches, not gradual change, make the link between quantitative and qualitative behavior, and form the basis for emergent phenomena. If this picture is correct for the real world, then we must accept instability and catastrophes as inevitable in biology, history, and economics. . . . Even though there are many more small events than large ones, most of the changes of the system are associated with large, catastrophic events. Self-organized criticality can be viewed as the theoretical justification of catastrophism." (*How Nature Works*, 31–32).

10   Bak, "The sandpile paradigm," ibid., 49–64.

11   On "complexism" as a "scientific ideology" see Christina Cogdell, *Towards a Living Architecture? Complexism and Biology in Generative Design* (Minneapolis, MN: University of Minnesota Press, 2018), 6, and passim.

12   Bak, *How Nature Works*, 183–198.

13   Warren Weaver, "A Quarter Century in the Natural Sciences," in *The Rockefeller Foundation Annual Report: 1958*; see in particular chapter 1, "Science and Complexity," 7–15 (an earlier version in chapter 1 of Weaver, *The Scientists Speak* [New York: Boni and Gaer, 1947]; edited and republished in Weaver, "Science and Complexity," *American Scientist* 36, no. 4 [October 1948]: 536–544). Weaver defines "problems of organized complexity" as "dealing simultaneously with a sizeable number of factors that are interrelated into an organic whole . . . in a complicated but not helter-skelter fashion" (14). Weaver never used the terms "emergence" or "system" in this context. He described the rise of a new science of "organized complexity" better suited to biology and to medical sciences, opposing it to the old science of "simplicity," or "two-variable science," used in classical mechanics, and to the science of "disorganized complexity," best represented by probability theory and statistics and applying, among others, to quantum mechanics, communication theory, and information theory. Weaver anticipated the extension of the new science of "organized complexity" from biology to behavioral and social sciences, but in his 1958 report he did not provide any insight on its tools and methods. Oddly, and perhaps significantly, in his 1948 article for *American Scientist* Weaver had argued that the new electronic computers developed during the war, due to their super-human computing performance, would "justify and inspire new methods of analysis applicable to these new problems of organized complexity" (Weaver, 1948: 541), but this passage was omitted from Weaver's better known Rockefeller Foundation Report of 1958.

14   Jane Jacobs was herself a grantee of the Rockefeller Foundation, and the amount of her grant figures in the same 1958 report as Weaver's essay. References to Weaver, and to Weaver's theory of "organized complexity," appear only in the last chapter of Jacobs's book, which reads as an addition or as a postface of sorts. Unlike Weaver, Jacobs refers explicitly to "systems of organized complexity." Jane Jacobs, *The Death and Life of Great American Cities* (New

York: Random House, 1992), 442; first published 1961.

15 "A complex system in Herbert A. Simon's definition includes 'a large number of parts that interact in a non-simple way.' The difficult whole in an architecture of complexity and contradiction includes multiplicity and diversity of elements in relationships that are inconsistent or among the weaker kinds perceptually." Robert Venturi, *Complexity and Contradiction in Architecture* (New York: Museum of Modern Art, 1966), 89. Venturi refers to Herbert A. Simon, "The Architecture of Complexity," *Proceedings of the American Philosophical Society* 106, no. 6 (1962): 467–82. See also Cogdell, *Toward a Living Architecture?*, 221.

16 Earlier versions of part of this section were published in Mario Carpo, "Digitally Intelligent Architecture has Little to do with Computers (and Even Less with their Intelligence)," *GTA Papers* 3 (2019): 112–120.

17 Using different metrics, Michael Wooldridge has recently calculated that a 2020 smartphone has "4,000 times more memory" than the most advanced mainframe computers of the late 1960s: Michael Wooldridge, *The Road to Conscious Machines: The Story of AI* (London: Pelican, 2020), 379.

18 Daniel Cardoso Llach, *Builders of the Vision: Software and the Imagination of Design* (New York; London: Routledge, 2015), 39, 49–72.

19 Ibid.

20 Oral communication from Philip Steadman (co-founder of the Centre for Land Use and Built Form Studies [LUBFS] at the School of Architecture of the University of Cambridge, 1967). There are some slightly different anecdotal traditions on what Sutherland would have actually shown to his British colleagues in 1963.

21 In 1968 SOM, one the biggest architectural firms of the time, and one of the few to own and operate a mainframe computer, presented some programs for cost estimates and building area calculations as new and ground-breaking research at a conference on "computing machinery." Cardoso Llach, *Builders of the Vision*, 23–24.

22 Alexandros-Andreas Kyrtsis, ed., *Constantinos A. Doxiadis: Texts, Design Drawings, Settlements* (Athens: Ikaros, 2006), 455.

23 Wiener explained that his own studies on the subject had been impelled by a war project on the self-correction of gun pointers aimed at airplanes with known or predictable trajectories. Norbert Wiener, *Cybernetics: or Control and Communication in the Animal and the Machine* (Cambridge, MA: MIT Press, 1948). The introduction is dated Mexico City, 1947. Introduction republished in the second, enlarged edition (1961), 1–30: 11.

24 Wiener, *Cybernetics*, 14, 19.

25 Ibid., 15, 23.

26 Nils J. Nilsson, *The Quest for Artificial Intelligence: A History of Ideas and Achievements* (New York: Cambridge University Press, 2010), 52–55. At the time of writing the best source of information on the Dartmouth workshop, seen by many as the act of foundation of Artificial Intelligence as a discipline, is still a remarkable Wikipedia entry, https://en.wikipedia.org/wiki/Dartmouth_workshop. We must assume that in this instance, contrary to its terms of service, but faithful to its spirit, Wikipedia serves as an aggregator of oral traditions, mostly contributed by the protagonists of the story being told, or by people who were close to them.

27 Marvin Minsky, "Steps Toward Artificial Intelligence," *Proceedings of the IRE* 49, no. 1 (1961): 8–30.

28 For an introduction to this discussion see Piero Scaruffi, *Intelligence is not Artificial* (self-pub., 2018), 19–23.

29 See the catalog of the exhibition: Jasia Reichardt, ed., *Cybernetic Serendipity: The Computer and the Arts* (London: Studio International, 1968). The digital scan of a photograph of Norbert Wiener (page 9) had a resolution of 100,000 black or white cells (i.e., the scan would have been 100 Kb in size). The caption describes the process and the technology used; the scan and print took sixteen hours of non-stop machine work.

30 Gordon Pask was the cybernetic consultant for Archigram's Instant City (1968) and the "cybernetic resident" in Cedric Price's Fun Palace (1963–67); he contributed to the eighth *Archigram* magazine, and he went on to collaborate (alongside John and Julia Frazer) on Price's Generator Project (1976–79). Usman Haque, "The Architectural Relevance of Gordon Pask," in Lucy Bullivant, ed., "4dsocial Interactive Design Environments," special issue (*AD Profile 188*), *Architectural Design 77*, no. 4 (July–August 2007), 54–61; Molly Wright Steenson, *Architectural Intelligence: How Designers and Architects Created the Digital Landscape* (Cambridge, MA: MIT Press, 2017), 156–175.

31 Steenson, *Architectural Intelligence*, 128, source not cited. Steenson adds that the Generator Project "actually showed how artificial intelligence could work in an architectural setting."

32 Nicholas Negroponte, *The Architecture Machine: Toward a More Human Environment* (Cambridge, MA: MIT Press, 1970), 104–105; Steenson, *Architectural Intelligence*, 185.

33 Steenson, *Architectural Intelligence*, 192–195; Scaruffi, *Intelligence is not Artificial*, 62–75.

34 John Walker [Autodesk founder], ed., *The Autodesk File: Bits of History, Words of Experience*, 3rd ed. (Thousand Oaks, CA: New Riders Publishing, 1989). Another edition (5th; 2017) is on Walker's website: http://www .fourmilab.ch/autofile/; Cardoso Llach, *Builders of the Vision*, 166.

35 Competition launched December 1969; results announced July 1971; construction started May 1972; building inaugurated January 31, 1977. Architects Richard Rogers, Renzo Piano; engineering: Edmund Happold, Peter Rice at Ove Arup and Partners.

36 Tschumi recalls that the first applications for the purchase of computers at Columbia's GSAPP were prepared in 1992 and 1993; the first machines were received in the summer of 1994, and the first "paperless studios" that actually used computers were taught by Stan Allen, Keller Easterling, Greg Lynn, and Scott Marble in the fall of 1994. Hani Rashid followed soon thereafter. See Bernard Tschumi, "The Making of a Generation: How the Paperless Studio Came About," in *When Is the Digital in Architecture*, ed. Andrew Goodhouse (Montreal: Canadian Centre for Architecture; Berlin: Sternberg Press, 2017), 407–419. The first paperless studios used Macintosh computers running Form Z, but there were also a few Silicon Graphics workstations running Softimage (an early animation software developed by Montreal filmmaker Daniel Langlois): Stan Allen, "The Paperless Studios in Context," in Goodhouse, *When is the Digital in Architecture*, 394. See also Nathalie Bredella, "The Knowledge Practices of the Paperless Studio," *GAM, Graz Architecture Magazine*, no. 10 (2014): 112–127.

37 Greg Lynn, ed., "Folding in Architecture," special issue (*AD Profile 102*), *Architectural Design 63*, nos. 3–4 (1993). Reprint, with new introductions by Greg Lynn and Mario Carpo (London: Wiley-Academy, 2004).

38 Mario Carpo, *The Alphabet and the Algorithm* (Cambridge, MA: MIT Press, 2011), 83–105.

39 Mario Carpo, *The Second Digital Turn* (Cambridge, MA: MIT Press, 2017), 189; Greg Lynn, ed., *Archaeology of the Digital* (Montreal: Canadian Centre for Architecture; Berlin: Sternberg Press, 2013), 55–57.

40 Carpo, *The Alphabet and the Algorithm*, 39–41.

41 Carpo, *The Second Digital Turn*, 132–140.

42 Starting with Greg Lynn, "Multiplicitous and Inorganic Bodies," *Assemblage* 19 (December 1992): 32–49, see in particular 35.

43 The exhibition *Breeding Architecture*, by Foreign Office Architects (Farshid Moussavi and Alejandro Zaera-Polo) was held at the Institute of Contemporary Art in London from November 29, 2003 to February 29, 2004 and was accompanied by the book *Phylogenesis: FOA's Ark*, ed. Michael Kubo, Albert Ferre, Farshid Moussavi, and Alejandro Zaera-Polo (Barcelona: Actar, 2003). The *Gen(H)ome Project* exhibition, curated by Kimberli Meyer, was held at the MAK Center at the Schindler House (Los Angeles) from October 29, 2006 to February 25, 2007; see the catalog of the exhibition, ed. Peter Noever (Los Angeles: MAK Center for Art and Architecture, 2006). See Cogdell, *Toward a Living Architecture?*, 117 and notes 14, 105.

44 John H. Holland, *Adaptation in Natural and Artificial Systems: an Introductory Analysis with Applications to Biology, Control, and Artificial Intelligence* (Ann Arbor: University of Michigan Press, 1975).

45 In order to fast-track the natural process of algorithmic self-improvement, so to speak, John Holland also famously claimed that we should breed algorithms as we breed horses, and the sinister connotations of this eugenic view of the world—not less sinister for being applied, metaphorically, to mathematical scripts—have been denounced by Christina Cogdell. John H. Holland, "Genetic Algorithms: Computer programs that 'evolve' in ways that resemble natural selection can solve complex problems even their creators do not fully understand," *Scientific American* 267, no. 1 (July 1992): 66–73; Cogdell, *Toward a Living Architecture?* (2018), 14–17, and passim.

46 John Frazer, *An Evolutionary Architecture* (London: Architectural Association, 1995).

47 Charles Jencks, *The Architecture of the Jumping Universe: A Polemic: How Complexity Science is Changing Architecture and Culture* (London: Academy Editions, 1995; rev. ed. London: Academy Editions, 1997; repr. Chichester: Wiley, 1998). See Georg Vrachliotis, "Popper's Mosquito Swarm: Architecture, Cybernetics and the Operationalization of Complexity," in *Complexity. Design strategy and world view*, ed. Andrea Gleiniger and Georg Vrachliotis (Basel: Birkhauser, 2008), 59–75.

48 Jencks, *The Architecture of the Jumping Universe*, 170–186.

49 Sanford Kwinter, "Not the Last Word," in Wes Jones (guest ed.), "Mech-In-Tecture: Reconsidering the Mechanical in the Electronic Era," *ANY: Architecture New York* 10 (February–March 1995): 61. Kwinter's column was titled "Far From Equilibrium" starting with *ANY* 11 (July–August 1995); some of the columns were republished in Kwinter, *Far From Equilibrium: Essays on Technology and Design Culture* (Barcelona: Actar, 2008).

50 Cogdell, *Toward a Living Architecture?* (2018), 28–30, 35–38, and passim.

51 Michael Hensel, Achim Menges, and Michael Weinstock, eds., "Emergence: Morphogenetic Design Strategies," special issue (*AD* Profile 169), *Architectural Design* 74, no. 3 (2004).

52 Michael Hensel, Achim Menges, and Michael Weinstock, *Emergent*

Technologies and Design: Towards a Biological Paradigm for Architecture (Abingdon: Routledge, 2010); Michael Weinstock, *The Architecture of Emergence: The Evolution of Form in Nature and Civilisation* (Chichester: Wiley, 2010).

53 Cogdell, *Toward a Living Architecture?* (2018), 39–41.

54 Stephen Wolfram, *A New Kind of Science* (Champaign, IL: Wolfram Media, 2002). See Carpo, *The Second Digital Turn*, 92–93, and footnotes.

55 Wolfram, *A New Kind of Science*, 876.

56 See note 13. Weaver was an engineer and a mathematician by training, a pioneer of automated translation, and chair of the Department of Mathematics of the University of Wisconsin before he was hired as Director for the Natural Sciences of the Rockefeller Foundation (1932 to 1959).

57 Alisa Andrasek, *The Invisibles*, Prague Biennale of 2003 (where Cellular Automata were used to generate sound, not images), published in Pia Ednie-Brown, "All-Over, Over-All: Biothing and Emergent Composition," in Mike Silver, ed., "Programming Cultures," special issue (*AD Profile 182*), *Architectural Design 76*, 4 (2006): 72–81. Philippe Morel's *Bolivar Chair* (2004): EZCT Architecture & Design Research, Philippe Morel with Hatem Hamda and Marc Schoenauer, *Studies on Optimization: Computational Chair Design Using Genetic Algorithms*, 2004; first presented at Archilab 2004 in Orléans and published in Philippe Morel, "Computational Intelligence: The Grid as a Post-Human Network," in Christopher Hight and Chris Perry, eds., "Collective Intelligence in Design," special issue (*AD Profile 183*), *Architectural Design 76*, no. 5 (2006): 100–103. A prototype and drawings of the "Bolivar" model are in the Centre Pompidou Architecture Collection. Neri Oxman, *Tropisms: Computing Theoretical Morphospaces*, research

project (MIT, 2006), published in Alisa Andrasek, Pia Ednie-Brown, "Continuum: A Self-Engineering Creature Culture," in Hight and Perry, "Collective Intelligence": 19–25, 20.

58 See notes 44, 45.

59 Claudia Pasquero and Marco Poletto, "Cities as Biological Computers," *arq: Architectural Research Quarterly 20*, no. 1 (March 2016): 10–19; Steven Shaviro, *Discognition* (London: Repeater, 2016). See in particular chapter 7, "Thinking Like a Slime Mold," 193–215. I am indebted to Claudia Pasquero for my induction into the world of slime molds.

60 A generalist, synoptic history of Artificial Intelligence does not exist to date, with the partial exceptions of Nilsson, *The Quest for Artificial Intelligence*; Scaruffi, *Intelligence is not Artificial*, and two recent books by Michael Wooldridge: *The Road to Conscious Machines*, and more particularly *A Brief History of Artificial Intelligence: What It Is, Where We Are, and Where We Are Going* (New York: Flatiron Books, 2021). My story builds in part on data gleaned from technical literature.

61 Marvin Minsky, "Steps Toward Artificial Intelligence." Minsky's paper was submitted in October 1960.

62 Marvin Minsky and Seymour Papert, *Perceptrons: An Introduction to Computational Geometry* (Cambridge, MA: MIT Press, 1969).

63 Carpo, *The Second Digital Turn*, 33–40.

64 Ibid., 38.

65 From different premises, and with more robust facts and figures to support its argument, this is also the conclusion of Phil Bernstein's recent *Machine Learning: Architecture in the Age of Artificial Intelligence* (London: RIBA Publishing, 2022), 46. Bernstein refers to Richard and Daniel Susskind's recent studies on the future of professional work: according to the Susskinds

automation will take over specific tasks, not entire jobs, and the tasks more prone to automation are those that entail explicit processes and goals that can be easily measured (as opposed to jobs based on "implicit knowledge"); Bernstein concludes that the adoption of AI and machine learning in the design professions will lead to the development of more intelligent design tools, not to the rise of a new breed of post-human designers. See Richard and Daniel Susskind, *The Future of the Professions: How Technology Will Transform the Work of Human Experts* (Oxford: Oxford University Press, 2015); Daniel Susskind, *A World Without Work: Technology, Automation, and How We Should Respond* (London: Penguin Books, 2020).

66 The technology was first outlined in a 2014 paper by Ian Goodfellow et al. ("Generative Adversarial Nets," *NIPS 2014, Proceedings of the 27th International Conference on Neural Information Processing Systems* 2 [December 2014]: 2672–2680). Applications of GAN to image processing followed soon thereafter, first for videogaming, then for artistic purposes; at the time of writing (2021) GAN are the main tool for the creation of AI-enabled art, or art generated by Artificial Intelligence. The use of GAN for visual "style transfer" was first theorized by Leon A. Gatys et al., in 2015–2016 (Leon A. Gatys, Alexander S. Ecker, and Matthias Bethge, "A Neural Algorithm of Artistic Style," *ArXiv*: 1508.06576 [August–September 2015]; Leon A. Gatys, Alexander S. Ecker, and Matthias Bethge, "Image Style Transfer Using Convolutional Neural Networks," *2016 IEEE Conference on Computer Vision and Pattern Recognition* [June 2016]: 2414–2423). Designers have since applied the same method to architectural images, and they are now adapting similar processes to the manipulation of three-dimensional objects: see Matias del Campo, Alexandra Carlson, and Sandra Manninger, "Towards Hallucinating Machines— Designing with Computational Vision," *International Journal of Architectural Computing* 19, no. 1 (March 2021): 88–103, and Matias del Campo, *Neural Architecture: Design and Artificial Intelligence* (Novato, CA: ORO Editions, 2022).

67 The core of Bellori's theory is in his 1664 lecture, "L'idea Del Pittore, Dello Scultore, e Dell'architetto Scelta dalle bellezze naturali superiore alla Natura" published as the preface of Giovan Pietro Bellori, *Le Vite de' Pittori, Scultori et Architetti moderni* (Rome: Mascardi, 1672); included, with vast commentaries, in Erwin Panofsky, *Idea. Ein Beitrag zur Begriffsgeschichte der älteren Kunsttheorie* (Leipzig-Berlin: Teubner, 1924); translated as "Bellori's Idea of the Painter, Sculptor, and Architect, Superior to Nature by Selection of Natural Beauties," in Panofsky, *Idea. A Concept in Art History*, trans. Joseph J. D. Peake (Columbia, SC: University of South Carolina Press, 1968), appendix 2, 154–174.

68 The phrase (in context: "to paint a good looking woman, I need to see many; but as there is a penury of good looking women, I have recourse to certain ideas that just spring to my mind," translation mine) is in Raphael's letter to Baldassare Castiglione, dated 1514, which has been variously attributed, including to Castiglione himself; first published 1559; first critical edition in Vincenzo Golzio, *Raffaello: nei documenti nelle testimonianze dei contemporanei e nella letteratura del suo secolo* (Vatican City: Pontificia Accademia Artistica dei Virtuosi al Pantheon, 1936). See Gabriele Morolli, *Le Belle Forme degli Edifici Antichi: Raffaello e il progetto del primo trattato rinascimentale sulle antichità di Roma* (Florence: Alinea, 1984), 44.

69 Carpo, *Metodo ed Ordini nella Teoria Architettonica dei Primi Moderni: Alberti, Raffaello, Serlio e Camillo* (Geneva: Droz, 1993), 121–131; "Citations, Method, and the Archaeology of Collage" (*Real Review*), 22–30; (*Scroope*) 112–119.

70 With the significant difference that, again, this common denominator is by definition unwritten and unknowable—as nobody knows which criteria are being singled out by the GAN learning process, and it appears that, based on the current state of the art, there may be no way to know. In a noteworthy precedent, in the late 1970s William Mitchell and George Stiny applied Stiny and Gips's shape grammar method (1971–72) to the design of an algorithm meant to capture and reproduce Palladio's "style." But the shape grammar method was, and still is, based on the explicit formalization of rules, hence the performance of the system was dependent on the sagacity of the analyst and on the pertinence of the rules (the "shape grammar") at the core of the algorithm; in that instance, the algorithm generated plans of Palladian-looking villas derived from the application of a set of rules inferred from the analysis of the ground plans in a corpus of Palladio's buildings. George Stiny, William J. Mitchell, "The Palladian Grammar," *Environment and Planning B: Planning and Design*, 5, no. 1 (June 1978): 5–18.

71 As the British political commentator (and complexity science vulgarizer) Matt Ridley recently put it in an interview with the *Wall Street Journal*: "If you think biological complexity can come about through unplanned emergence and not need an intelligent designer, then why would you think human society needs an 'intelligent government'?" (Tunku Varadarajan, "How Science Lost the Public's Trust," *The Wall Street Journal*, print edition, July 24, 2021).

Ridley, a conservative member of the House of Lords in the UK, is a noted libertarian, Brexiteer, and occasional climate change denier.

Chapter 4

1 For an earlier discussion of some of the topics under review in this chapter see: Mario Carpo, "Citations, Method, and the Archaeology of Collage," *Real Review* 7 (2018): 22–30, republished in *Scroope, Cambridge Architectural Journal*, 28 (2019): 112–119; "On the Post-Human Charm of Chunky Beauty," in *Beauty Matters*, Tallinn Architecture Biennale 2019, ed. Yael Reisner (Tallin: Estonian Centre for Architecture, 2019), 92–103; "The Challenger. Dear Daniel and Gilles . . . ," *Flat Out* 4 (2020): 28–30; "PoMo, Collage and Citation. Notes Towards an Etiology of Chunkiness," in Owen Hopkins and Erin Mckellar, eds., "Multiform, Architecture in an Age of Transition," special issue (*AD Profile 269*), *Architectural Design* 91, no. 1 (2021): 18–25; for most of the classical sources see Carpo, *Metodo ed Ordini nella Teoria Architettonica dei Primi Moderni: Alberti, Raffaello, Serlio e Camillo* (Geneva: Droz, 1993).

2 Mario Carpo, "Breaking the Curve. Big Data and Digital Design," *Artforum* 52, no. 6 (2014): 168–173; *The Second Digital Turn* (Cambridge, MA: MIT Press, 2017), 9–40, 70–79.

3 Pliny, *Naturalis Historia*, Book 35, 64; Cicero, *De Inventione*, Book 2, 1; Xenophon, Ἀπομνημονεύματα (*Memorabilia*), 3.10.2 (where the topos is attributed to Socrates in conversation with Parrhasius, Zeuxis's rival). Zeuxis's anecdote will remain central to classicist aesthetics till Raphael (1514), Bellori (1664–1672) and beyond (see here chapter 3, notes 68–69).

4 The earliest probes into the epistemological implications of the Zeuxis story are in the letters *On*

185

Imitation between Giovan Francesco Pico della Mirandola and Pietro Bembo (1512–13; first printed 1518; translated into English in Scott, 1910; first critical edition in Santangelo, 1954), and in Giulio Camillo (also called Giulio Camillo Delminio), *L'idea dell'eloquenza* (ca. 1530; first published 1983: Lina Bolzoni, "L'idea dell'eloquenza. Un'orazione inedita di Giulio Camillo," *Rinascimento*, n.s., 23 [1983]: 125–166; and Bolzoni, *Il teatro della memoria, studi su Giulio Camillo* [Padua: Liviana, 1984], 107–127). See Izora Scott, *Controversies Over the Imitation of Cicero as a Model for Style and Some Phases of Their Influence on the Schools of the Renaissance* (New York: Columbia University Teachers College, 1910); Giorgio Santangelo, *Le epistole "De imitatione" di Giovan Francesco Pico della Mirandola e Pietro Bembo* (Florence: Olschki, 1954); Eugenio Battisti, "Il concetto di imitazione nel Cinquecento da Raffaello a Michelangelo," *Commentari* 7, no. 2 (1956), republished in Battisti, *Rinascimento e Barocco* (Turin: Einaudi, 1960), 175–216; Thomas M. Greene, *The Light in Troy: Imitation and Discovery in Renaissance Poetry* (New Haven, CT: Yale University Press, 1982), 175; Carpo, *Metodo ed Ordini*, 34–39, 47–52; Colin Burrow, *Imitating Authors: Plato to Futurity* (New York: Oxford University Press, 2019), 169–220. See also chapter 3, notes 68–70.

5 Desiderius Erasmus, *Dialogus cui titulus Ciceronianus sive de optimo genere dicendi* (Basel: Froben, and Paris: Simon de Colines, 1528); *Ciceronianus: or, a Dialogue on the Best Style of Speaking*, trans. Izora Scott (New York: Columbia University Teachers College, 1908); also in Scott, *Controversies Over the Imitation of Cicero* (1910). On "cut-and-paste" citations as a form of literary imitation in early modern rhetoric see Terence Cave, *The Cornucopian Text: Problems of Writing in the French Renaissance* (Oxford: Clarendon Press, 1979); Antoine Compagnon, *La seconde main, ou le travail de la citation* (Paris: éditions du Seuil, 1979); Marc Fumaroli, *L'âge de l'éloquence. Rhétorique et "res litteraria" de la Renaissance au seuil de l'époque classique* (Geneva: Droz, 1980): see in particular 95–98, on the rise of the early modern "rhetoric of citations" (*rhétorique des citations*).

6 Giulio Camillo, *L'idea del theatro* (Florence: Torrentino, 1550), written ca.1544.

7 Giulio Camillo, *L'idea dell'eloquenza* (ca. 1530); first published 1983: see note 3.

8 Mario Carpo, "Ancora su Serlio e Delminio. La Teoria Architettonica, il Metodo e la Riforma dell'Imitazione," in *Sebastiano Serlio*, ed. Christof Thönes (Milan: Electa, 1989), 110–114; Carpo, *Metodo ed Ordini*, 65–138; Carpo, "Citations, Method, and the Archaeology of Collage," (*Real Review*) 22–30; (*Scroope*) 112–119.

9 Starting with Lomazzo's famous tirade: with his books, Lomazzo lamented, Serlio "has made more dog-catchers into architects than he has hairs in his beard." Giovan Paolo Lomazzo, *Trattato dell'Arte della Pittura* (Milan: P. G. Pontio, 1584), 407. See Mario Carpo, *Architecture in the Age of Printing: Orality, Writing, Typography and Printed Images in the History of Architectural Theory* (Cambridge, MA: MIT Press, 2001), 117–118.

10 Manfredo Tafuri, "Ipotesi sulla religiosità di Sebastiano Serlio," in *Sebastiano Serlio*, ed. Christof Thönes (Milan: Electa, 1989), 57–66; Tafuri, *Venezia e il Rinascimento* (Turin: Einaudi, 1985), 90–112; English translation *Venice and the Renaissance*, trans. Jessica Levine (Cambridge, MA: MIT Press, 1989), 51–81; Mario Carpo, *La Maschera e il Modello: Teoria Architettonica ed Evangelismo nell'Extraordinario Libro di Sebastiano Serlio* (Milan: Jaca Book, 1993).

11 William M. Ivins, Jr., *Prints and Visual Communication* (Cambridge, MA: Harvard University Press, 1953; reprint Cambridge, MA: MIT Press, 1992), 28, 43; Arthur M. Hind, *An Introduction to a History of Woodcut, with a Detailed Survey of Work Done in the Fifteenth Century* (London: Constable and Co., 1935), 284, 498, 669.

12 Carpo, *La Maschera e il Modello*, 85–105; *Architecture in the Age of Printing*, 180; Sabine Frommel, *Sebastiano Serlio architetto* (Milan: Electa, 1998), 77–78.

13 John Summerson, *Architecture in Britain, 1530–1830* (London: Penguin Books, 1953), 97–105.

14 Owen Hopkins, *From the Shadows: The Architecture and Afterlife of Nicholas Hawksmoor* (London: Reaktion Books, 2015).

15 Clement Greenberg, "Collage," in *Art and Culture: Critical Essays* (Boston: Beacon Press, 1961), 70–84.

16 Colin Rowe, "The Mathematics of the Ideal Villa: Palladio and Le Corbusier Compared," *Architectural Review* 101, no. 603 (March 1947): 101–104, republished as "The Mathematics of the Ideal Villa" in Rowe, *The Mathematics of the Ideal Villa and Other Essays* (Cambridge, MA: MIT Press, 1976), 1–17. Rowe's article was published one year before he submitted his thesis for a master's degree in the history of art as Wittkower's only student at the Warburg Institute ("The Theoretical Drawings of Inigo Jones: Their Sources and Scope," 1948, unpublished): see Mollie Claypool, "The Consequences of Dialogue and the Virgilian Nostalgia of Colin Rowe," *Architecture and Culture* 4, no. 3 (2016): 359–367. Wittkower's own take on Palladio's proportional systems would be published in his *Architectural Principles in the Age of Humanism* (London: The Warburg Institute, 1949).

17 Colin Rowe and Fred Koetter, *Collage City* (Cambridge, MA: MIT Press, 1983), 143; first published 1978.

18 The foundational texts of the theory of intertextuality are Kristeva's article "Bakhtine, le mot, le dialogue et le roman," *Critique* 23, no. 239 (1967): 438–465, republished in Kristeva, *Séméiôtiké: recherches pour une sémanalyse* (Paris: Seuil, 1969), 82–112 (where the article is dated 1966); and Kristeva's contribution to the Colloque de Cluny of 1968, published in *Théorie d'Ensemble*, ed. Roland Barthes, Jean-Louis Baudry, and Jacques Derrida [Collection *Tel Quel*] (Paris: Seuil, 1969); see in particular the chapter "L'intertextualité, le texte comme idéologème," 312–318. The official consecration of Kristeva's theory came with Roland Barthes's entry "Texte, théorie du," in the 1974 edition of the *Encyclopaedia Universalis*, where Barthes offered the now famous aphorism "tout texte est un tissu nouveau de citations révolues." See vol. 15, 1013–1017.

   This overarching and often militant interest in intertextuality as a compositional device was the common background for a number of scholarly studies on the history and theory of cut-and-paste citations, both literary and visual, that flourished in the late 1970s and in the 1980s: see for example the works of Terence Cave, Antoine Compagnon, or Marc Fumaroli, cited here in note 4, above, all published in 1979–80. Above and beyond the postmodern partiality for collages, including the historicist "collage city" advocated by Colin Rowe, the intertextual furor of the 1970s and early 1980s has left many a trace in architectural theory and also in the architectural historiography of the time: see, for example, various attempts at presenting Leon Battista Alberti as an early champion of visual and literary intertextuality, or "citationism," notably in *Leon Battista Alberti*, ed. Joseph Rykwert and Anne Engel (Milan: Electa, and Ivrea: Olivetti, 1994), 24–27.

19 In David Lodge's academic lampoon *Small World* (first published 1984) the fictional literary critic Morris Zapp uses the catchphrase "every decoding is another encoding" to show off his discovery of Derridean Deconstructionism, but also confessing to his growing disillusion and cynicism—if all interpretation is rigged, all communication is pointless, and all mediation by words is doomed, human society reverts to the state of nature, and to the law of the jungle.

20 In the fall term of 2013 Eisenman's studio at the Yale School of Architecture was titled "A Project of Aggregation"; the aim of the studio (Yale SoA 1105a), as outlined in the syllabus, was to "redefine the term 'aggregation' as a critique of the digital, which has become synonymous with homogeneous continuous space." In November that year, during a conference which took place in Belgrade, Serbia, and of which the transactions are now published, Eisenman claimed that he had already used the term "aggregation" in the 1980s, in a similar spirit and with a similar meaning—back then to champion "heterogeneous space" against classical composition, as now to plead for heterogeneity against the smooth spaces created by digital parametricism. Even though there is a consistent oral tradition suggesting that the term "aggregation" has long been used by formalist architectural thinkers and design educators, particularly in the context of the theory of part-to-whole relationships, I can't provide documental evidence of the occurrence of the term in Eisenman's writing or teaching before 2010–2011. Nor can Eisenman himself, with whom I recently (October 2021) discussed the matter. The brief of Eisenman's 2013 studio is published in *Retrospecta*, a publication of the Yale School of Architecture, vol. 37, 2013–14: see Michael Jasper, "Composing the Whole Thing: Design Research in Peter Eisenman's Recent Studio Teaching," *Annual Design Research* (Sydney: University of Sydney, 2018), 687–698; for the transactions of the Belgrade Conference ("Issues? Concerning the Project of Peter Eisenman," University of Belgrade, November 11–12, 2013), see *Peter Eisenman in Dialogue with Architects and Philosophers*, ed. Vladan Djokić and Petar Bojanić (Sesto San Giovanni: Mimesis Edizioni, 2017); for Peter Eisenman on "aggregation," 56.

21 Peter Eisenman, "The Challenger. Dear Mario . . ." *Flat Out* 3 (2019): 81–82. See also Carpo, "The Challenger. Dear Daniel and Gilles . . . ," *Flat Out* 4 (2020): 28–30.

22 See Gilles Retsin, ed., "Discrete. Reappraising the Digital in Architecture," special issue (*AD* Profile 258), *Architectural Design* 89, no. 2 (2019); and Mollie Claypool, Manuel Jimenez Garcia, Gilles Retsin, and Vicente Soler, *Robotic Building: Architecture in the Age of Automation* (Munich: Detail, 2019).

23 The expression appears to derive from an exchange between Robert E. Somol and Pier Vittorio Aureli, circa 2004: see Somol, "12 Reasons to Get Back into Shape," in *Content*, ed. Rem Koolhaas et al. (Cologne: Taschen, 2004), 86–87; and Aureli, "Who's Afraid of the Form-Object," *Log* 3 (Fall 2004): 29–36.

24 Wolfgang Köhler, *Gestalt Psychology* (New York: Liveright, 1929), 258.

25 See, for example, the representation of human suffering conveyed by load-bearing columns built in the shape of feminine or masculine bodies (caryatids, telamons) in *De architectura*, 1.1.4–6.

26 Digital collages were pioneered in the early 2000s by Dogma and OFFICE Kersten Geers David van Severen, and they were critically

endorsed in a noted 2017 article by Sam Jacob, who saw them as the core representational tool and visual trope of a new "post-digital" wave. Both Pier Vittorio Aureli, of Dogma, and Jacob, saw digital collages (i.e., the perverted use of digital tools to imitate the semblances of a mechanical cut-and-paste process) as a reaction against the smooth hyperrealism of early digital renderings; Aureli's and Jacob's revival of the collage aesthetics was hence meant to represent their ideological rejection of post-mechanical technologies, and their nostalgic predilection for vintage, mechanical ones. To be noted that when, circa 2011–13, Peter Eisenman revived his Deconstructivist theory of formalist "aggregation," he claimed to be motivated by a similarly militant, anti-digital purpose: see here note 19, above. Neoluddite arguments are sometimes invoked by other "post-digital" designers, often resorting to collages of classicist citations, in the tradition of Colin Rowe and historicist postmodernism. See: Dogma, "Architecture Refuses," GIZMO, December 12, 2008 (http://www.gizmoweb.org/2008/12/dogma-architecture-refuses); Sam Jacob, "Architecture Enters the Age of Post-Digital Drawing," *Metropolis*, March 21, 2017 (https://www.metropolismag.com/architecture/architecture-enters-age-post-digital-drawing); Owen Hopkins, Erin McKellar, ed., "Multiform, Architecture in an Age of Transition," special issue (*AD* Profile 269), *Architectural Design* 91, no. 1 (2021). I owe some of these references to the master's thesis of my student Alejandro Carrasco Hidalgo (Bartlett School of Architecture, Master's in Architectural History, 2020–21).

27 "Quod visum perturbat," as in Reyner Banham's famous definition of New Brutalism in 1955. Banham,

"The New Brutalism," *Architectural Review* 118, no. 708 (December 1955): 354–361.

Epilogue

1 Florian Cramer, "What Is 'Post-Digital'?," in *Postdigital Aesthetics: Art, Computation and Design*, ed. David M. Berry and Michael Dieter (London: Palgrave Macmillan, 2015), 12–26, https://doi.org/10.1057/9781137437204_2. First published online in the "Transactions of the Post-Digital Research Conference/Workshop," Kunsthal Aarhus, DK, 7–9 October 2013 (http://post-digital.projects.cavi.au.dk/?p=599) and in *APRJA* 3, no. 1 (2014), https://doi.org/10.7146/aprja.v3i1.116068.

2 Sam Jacob, "Architecture Enters the Age of Post-Digital Drawing." See above, chapter 4, note 26 *Metropolis*, March 21, 2017, https://www.metropolismag.com/architecture/architecture-enters-age-post-digital-drawing.

Index

197

von Neumann, John, 96, 109
Vrachliotis, Georg, 182n47

Waldman, Jonathan, 176n75
Walker, John, 181n34. *See also* Autodesk
Warburg Institute, 142, 187n16
Watanabe, Hiroshi, 178n89
Weaver, Warren, 93, 109, 179nn13–14,
    183n56
    Rockefeller Foundation Report
        (1958), 93, 109, 179n13
Webb, Clifton, 174n35
Weinstock, Michael, 107, 182n51–52.
        See also Emergence and Design
        Group, Architectural Association
        School of Architecture (AA);
        Hensel, Michael; Menges, Achim
Westinghouse, 54
    automatic air brake (patent), 54
Whitney, Eli, 6. *See also* Colt, Samuel
Wiener, Norbert, 53, 94–97, 100, 175n65,
        180nn23–25, 181n29. *See also*
        Rosenblueth, Arturo
    *Cybernetics: Or Control and
        Communication in the Animal and the
        Machine*, 95, 180nn23–25
Wikipedia, 180n26
Wiley Marshall, Margaret,
        174–175nn52–53
Williams, Alex, 170n27
Willmann, Jan, 176nn74–76, 177nn80, 84
Wittkower, Rudolf, 142, 187n16
Wolfram, Stephen, 108–109, 183nn54–55
    Mathematica, 108
    New Kind of Science, A, 108, 183nn54–55
Womack, James, 165n8
Wooldridge, Michael, 180n17, 183n60
Wyss Institute, Harvard University, 71
    Kilobots, 71

Xenophon, 185n3

Yale School of Architecture, 188n20
    "A Project of Aggregation" (studio),
        188n20
Young, Thomas, 83
Yuan, Philip F., 172n6, 177n83

Zaera-Polo, Alejandro, 168n19, 182n43.
        See also Foreign Office Architects
        (FOA)
Zeldman, Maurice I., 175n66

Zeuxis, 132–133, 185nn3–4
Zinser, Michael, 167n16
Zipf, George Kingsley, 86, 88–90, 178n6
    Zipf's law, 86, 88–90, 178n6
Zone Books, 107
Zoom, 2, 27, 30

The MIT Press would like to thank the anonymous peer reviewers who provided comments on drafts of this book. The generous work of academic experts is essential for establishing the authority and quality of our publications. We acknowledge with gratitude the contributions of these otherwise uncredited readers.

This book was set in Lexicon by Jen Jackowitz.
Printed and bound in the United States of America.

Library of Congress Cataloging-in-Publication Data

Names: Carpo, Mario, author.
Title: Being post-digital : design and automation at the end of modernity / Mario Carpo.
Description: Cambridge, Massachusetts : The MIT Press, [2023] | Includes bibliographical references and index.
Identifiers: LCCN 2022014693 (print) | LCCN 2022014694 (ebook) | ISBN 9780262545150 (paperback) | ISBN 9780262373395 (epub) | ISBN 9780262373401 (pdf)
Subjects: LCSH: Architecture—Data processing. | Architecture—Computer-aided design.
Classification: LCC NA2728 .C375 2023 (print) | LCC NA2728 (ebook) | DDC 720.285—dc23/eng/20220720
LC record available at https://lccn.loc.gov/2022014693
LC ebook record available at https://lccn.loc.gov/2022014694

10   9   8   7   6   5   4   3   2   1